College Preparatory Mathematics 3
(Algebra 2)

Parent Guide with Review
Second Edition

Parent Guide Editor:

Karen Wootton
CPM Educational Program

Program Directors:

Tom Sallee
Department of Mathematics
University of California, Davis
Judith Kysh
The CRESS Center
University of California, Davis
Elaine Kasimatis
Department of Mathematics
California State University, Sacramento
Brian Hoey
CPM Educational Program

Technical Assistance:

Bethany Sorbello
CPM Educational Program
Thu Pham
The CRESS Center
University of California, Davis

Illustrator:

Kelly DeMartini

Textbook Editors:

Steve Simon
Lick-Wilmerding High
Robert Petersen
Sacramento High
Judith Kysh
The CRESS Center
University of California, Davis
Sally Keszler
Bear Creek High

v. 5.1

Published by

CPM Educational Program
1233 Noonan Drive
Sacramento, CA 95822-2569
E-mail: cpm@cpm.org

3 4 5 6 7 8 9   03 02 01 00

Printed in the United States of America   ISBN 1-885145-66-7

Table of Contents

To The Student

WELCOME!

I presume you are enrolled in an Algebra 2 class known as CPM Mathematics 3, and since you are reading this, you probably have some questions. First be assured that you are taking the first step toward better understanding by referring to this guide. Second, you can be successful in this (and all) math courses you take. You have probably read in the text that the authors have written this course with the goal being your understanding and success. Your teacher is probably using this text because he/she feels that this is the best book to promote learning and understanding. So take this to be true: there is a long list of people on your side who want you to do well in this course.

I'm also on that list. My name is Karen Wootton and I've been a teacher since 1984. I believe a strong education for everyone is the key to a strong future. I want you to do well in your classes and I've written this guide to help you gain a better understanding of the course. I've included the key pieces of information you need to be successful.

In this guide you won't find all of the answers to all of the problems. What you will find is help: suggestions for how to approach problems, what to think about when doing problems, some solutions, and some additional problems for you to try. The problems and questions I pose may not be identical to problems in the book, but that's okay. We don't want you to just be good at repeating how to do one type of problem. The problems I give you are realistic and appropriate. I taught Algebra 2 for several years so I know what you have done and what you will be doing. This guide is a good source book for you.

The next step is the tricky one: now you must use the information here and in your text book, along with your brain, and **think**. "What's so tricky about that?" you might ask. The problem is that many students don't realize that thinking, whether it is about solutions or procedures, is work. Just as you wouldn't expect a basketball player to improve by only watching the games from the sidelines, you can't expect to improve by sitting back and watching your math class. You must be actively involved. One of the best ways to help you be involved is by having you work in study teams. In many classes the teacher stands in front of the students and does most of the talking and most of the work. In this class, you will work on guided investigations--somewhat like "math labs"--that will help you understand the ideas of Algebra 2. Your teacher will take on the role of your coach, sometimes talking to the entire class, sometimes talking with the study teams, and sometimes working with students individually. This guide will also be a coach for the times when your study teammates or teacher are not available. I will offer suggestions to improve your mathematical abilities and ask you questions to prompt new ideas from you.

In using this guide you will find that some examples are easy and some are hard. **The important thing is that you try each one before reading the solution.** Even if you feel the example is too hard, you must try it first. Also, I am assuming that you have done or have attempted every problem your teacher assigns. I will frequently refer to a method or strategy used in a previous problem and it will be important that you know what I'm talking about.

This guide follows the book unit by unit, but not necessarily day by day or problem by problem. Occasionally I might skip around within a unit to show similarities or connections between problems. At the end of each section you will find more problems like those in the unit for you to use for additional practice. There are also some SAT-type math questions to try. Since these are based on actual tests, they may look different than what you have been doing in class and your textbook. Answers to all of these problems are at the end of the guide.

Let's get started!

With expectations of success,
Karen

UNIT 1

Sharpening Pencils

EXPLORING FUNCTIONS

UNIT 1
EXPLORING FUNCTIONS

Many of the ideas covered in Unit 1 should be review, but that doesn't mean that all the problems will be easy. First you must remember the idea or concept that the problem is covering. Second, the idea or concept will probably be in a new situation. Part of your task is to see how the problems are similar to ones you have seen before. A good problem solving strategy is to relate the new problem to a similar, hopefully simpler problem. If you can solve the similar problem, that solution can help you solve the new problem.

To begin with, let's take a quick look at solving equations. You will be solving many equations throughout the year, so don't think they are going to go away! Luckily for you, this will give you many opportunities to practice. Also, one of the great things about solving equations is that you can **always** check your answer. I'm certain that if your teacher gave you the answers to the test you were taking you would definitely check to see if your answers were correct. Similarly, when you solve an equation, you must take the time to check your solution. Not only will you see if your answer is correct, but when you check, you are also getting additional practice with equations, expressions, and order of operations.

Below are four equations to solve: a linear equation, solving implicitly, two equations with two unknowns (also called simultaneous equations), and a quadratic equation. Copy the equations on another sheet of paper and solve them **before** you look at the solution. Check your answer yourself before you look at the solution. Then you will know whether or not your answer is correct before you even look.

1. Solve for x.
$$6(2x - 7) = 4x - 6$$

Solution:
$$\begin{aligned} 6(2x - 7) &= 4x - 6 \\ 12x - 42 &= 4x - 6 \\ 8x &= 36 \\ x &= 4.5 \end{aligned}$$

Check:
$$\begin{aligned} 6(2(4.5) - 7) &\stackrel{?}{=} 4(4.5) - 6 \\ 6(9 - 7) &\stackrel{?}{=} 18 - 6 \\ 6(2) &= 12 \ \checkmark \end{aligned}$$

2. Solve for y in terms of x.
$$5x - 6y = 10$$

Solution:
$$\begin{aligned} 5x - 6y &= 10 \\ -6y &= -5x + 10 \\ y &= \frac{-5x + 10}{-6} \\ y &= \frac{-5x}{-6} + \frac{10}{-6} \\ y &= \frac{5x}{6} - \frac{5}{3} \end{aligned}$$

Check:
$$\begin{aligned} 5x - 6\left(\frac{5x}{6} - \frac{5}{3}\right) &\stackrel{?}{=} 10 \\ 5x - 5x + 2(5) &\stackrel{?}{=} 10 \\ 10 &= 10 \ \checkmark \end{aligned}$$

3. Solve for x and y.

$$3x - 4y = 2$$
$$y = \frac{1}{4}x + 3$$

Solution:

$$3x - 4\left(\frac{1}{4}x + 3\right) = 2$$
$$3x - x - 12 = 2$$
$$2x - 12 = 2$$
$$2x = 14$$
$$x = 7$$
$$y = \frac{1}{4}(7) + 3$$
$$y = 4.75$$

Check:

$$3(7) - 4(4.75) \stackrel{?}{=} 2$$
$$21 - 19 = 2 \ \surd$$

And,

$$4.75 \stackrel{?}{=} \frac{1}{4}(7) + 3$$
$$4.75 = 1.75 + 3 \ \surd$$

4. Solve for x.

$$12x^2 - 22x + 8 = 0$$

Solution:

$12x^2 - 22x + 8 = 0$

Using the Quadratic Formula,
$a = 12$, $b = -22$, and $c = 8$

$$x = \frac{22 \pm \sqrt{22^2 - 4(12)(8)}}{2(12)}$$
$$x = \frac{22 \pm \sqrt{484 - 384}}{24}$$
$$x = \frac{22 \pm \sqrt{100}}{24}$$
$$x = \frac{22 \pm 10}{24}$$
$$x = \frac{22 + 10}{24}$$

And,

$$x = \frac{22 - 10}{24}$$

So,

$$x = \frac{32}{24} \text{ and } \frac{12}{24}$$

or $x = \frac{4}{3}$ and $\frac{1}{2}$

Check:

$$12x^2 - 22x + 8 = 0$$
$$12\left(\frac{4}{3}\right)^2 - 22\left(\frac{4}{3}\right) + 8 \stackrel{?}{=} 0$$
$$12\left(\frac{16}{9}\right) - \frac{88}{3} + 8 \stackrel{?}{=} 0$$
$$\frac{64}{3} - \frac{88}{3} + 8 \stackrel{?}{=} 0$$
$$-8 + 8 = 0 \ \surd$$

And,

$$12\left(\frac{1}{2}\right)^2 - 22\left(\frac{1}{2}\right) + 8 \stackrel{?}{=} 0$$
$$12\left(\frac{1}{4}\right) - 11 + 8 \stackrel{?}{=} 0$$
$$3 - 11 + 8 = 0 \ \surd$$

Note: I will place this border before each example and again at the end of the solution.

When you solved these problems you may not have shown every step that I did. That is fine if you did the problem correctly. But if you find that you skipped various steps and you did not do the problem correctly, then you should start showing more work. **Showing your work is an excellent way to show your teacher how much you know.**

The first problem should have been familiar to you. You began solving equations like that in Algebra 1. Always check to be sure you are doing these correctly. When you solve an equation in terms of other variables, as in number 2, we say you are **solving implicitly.** You do these sorts of problems the same as the first example; you just can't simplify the expression as much. To solve the two equations in number 3, I used the substitution method. You might have used another method. In Unit 2 you will have lots of practice with simultaneous equations and we will review the other solution methods as well. To solve the quadratic equation of number 4, I used the Quadratic Formula. I'm sure you recall that some quadratic equations can be solved by factoring; as a matter of fact, this one can. But remember that the Quadratic Formula works on **every** quadratic equation. A good rule of thumb is to always try to factor a quadratic expression first, but if you can't come up with the factors within a few seconds, then stop and use the Quadratic Formula. Make sure you have the formula along with an example in your tool kit. Even if you use the formula often, write it down. You never know when your mind might draw a blank.

In Unit 1 you will also find several problems asking you to find the error in a solution. The best way to do these problems is for you to cover up the steps of the solution and do the problem yourself, then compare the solutions. Many students, when reading mathematics, gloss over the details. In this type of problem, however, the details are the last thing you want to miss! Try this one.

5. Find the error in the solution. Identify the error and show how to do the problem correctly.

$$\begin{aligned} -4(x-3)+12x &= 9-7(9-2x) \\ -4x+12+12x &= 9-63-14x \\ 8x+12 &= -49-14x \\ 22x &= -66 \\ x &= -3 \end{aligned}$$

Solution.

This is the perfect example of why you want to check your work. If a student of mine had written this solution I would see that the student clearly knows how to solve equations, but that in this problem s/he has made an error. The error is probably the result of a careless mistake. That mistake would have been found when the student tried to check the solution. Sometimes when students get a "nice" answer such as this (-3 rather than something like 6.282), they assume that means they must be correct and don't bother checking. But in this problem, that's not the case. Did you find the error? The negative sign on the right side of the equality was not distributed correctly. Here is the correct solution.

$$\begin{aligned} -4(x-3)+12x &= 9-7(9-2x) \\ -4x+12+12x &= 9-63+14x \\ 8x+12 &= -49+14x \\ -6x &= -66 \\ x &= 11 \end{aligned}$$

Check:

$$\begin{aligned} -4(11-3)+12(11) &\stackrel{?}{=} 9-7(9-2(11)) \\ -4(8)+132 &\stackrel{?}{=} 9-7(9-22) \\ -32+132 &\stackrel{?}{=} 9-7(-13) \\ 100 &= 9+91 \quad \surd \end{aligned}$$

I realize that checking a solution can be messy sometimes. Certainly checking the two solutions to number 4 earlier was a long, messy process. You have a tool, however, to make this easier: your graphing calculator. Maybe your calculator isn't programmed to solve quadratic equations, but you can use it to quickly check the solution. Simply type in the left side of the equality, substituting the value you would like to check:

12(4/3)^2 - 22(4/3) + 8

and press enter. If you get 0, it checks! Most graphing calculators have a way to recall the last line. If yours does, recall the same expression, and then edit it with the second value of x to check:

12(1/2)^2 - 22(1/2) + 8

Press enter to check if the value is 0. Be sure you have a complete "how to" section in your tool kit about your calculator. Your calculator can do many things, so keep track of them in an organized manner.

Using your graphing calculator to check as above is similar to evaluating expressions in function notation. There are several problems in the this first unit (and there will be several more throughout the year) similar to this next one.

6. If $f(x) = 3x^2 - 7x + 12$, find
a) f(4) **b) f(-2)** **c) f(1.2)**

Solution.
This is the same as asking "Find the value of the expression $3x^2 - 7x + 12$ if $x = 4$." Using function notation, f(x), is just another way of naming the function. It could be called g(x), u(x), or even Bob(x). Writing the "(x)" emphasizes that x is the input. We could also write t(n) if we want n to be the input. Finding these values is easily done on the graphing calculator or by hand.

a)
$$\begin{aligned} f(4) &= 3(4)^2 - 7(4) + 12 \\ &= 3(16) - 28 + 12 \\ &= 48 - 28 + 12 \\ &= 32 \end{aligned}$$

b)
$$\begin{aligned} f(-2) &= 3(-2)^2 - 7(-2) + 12 \\ &= 3(4) + 14 + 12 \\ &= 12 + 14 + 12 \\ &= 38 \end{aligned}$$

c)
$$\begin{aligned} f(1.2) &= 3(1.2)^2 - 7(1.2) + 12 \\ &= 3(1.44) - 8.4 + 12 \\ &= 4.32 - 8.4 + 12 \\ &= 7.92 \end{aligned}$$

If you do these by hand, be sure that you follow the order of operations. Have them written on your tool kit in case you ever draw a blank: parentheses, exponents, multiplication and division (left to right), and addition and subtraction (left to right). Oftentimes students forget the "left to right" with multiplication, division, addition and subtraction, but it is necessary. Multiplication does not necessarily come before division, nor does addition necessarily come before subtraction. Part (a) above illustrates this in the second to last line. We do this problem left to right: first do the subtraction, then add. **Do not** add first and then subtract. If you did, you would get the wrong answer of 8. If you find that you miss many of these problems or other problems involving order of operations, then use your graphing calculator whenever you can -- either to check or to find the answer.

Although you can use your graphing calculator for evaluating and checking, the main reason for using a graphing calculator is to graph. To see how comfortable you are graphing with your graphing calculator, try this. Set the window at $-10 \le x \le 10$ and $-8 \le y \le 8$ and graph $y = x^2 + 16x + 52$. What do you see? Is this a complete graph? Remember that a complete graph is one in which you can see the shape of the graph completely and any part of the graph not on the screen is predictable based on what is on the screen. Since the equation has an exponent of 2 as its highest exponent, the graph of this equation is a parabola.

This is one example of why a graphing calculator alone is not enough. We must also use our brains. If we didn't have the knowledge that the graph would be a parabola, we might assume that the graph is just a straight line. What should you do so that you see the parabola in the window? Change the window until you find a suitable frame. One possibility is $-20 \le x \le 0$ and $-15 \le y \le 1$. You can probably do better. If you ever produce a graph on your calculator that just doesn't seem right, check the window. Vary it and see how your graph changes. Another thing to check if a graph seems peculiar is whether or not you typed the equation in correctly. Consider these two equations.

Graph each of these on your graphing calculator. Be sure to show a complete graph.

7. $y = \dfrac{1}{x + 2}$

8. $y = \dfrac{x + 2}{x - 3}$

Remember! Try these before looking at the solution!

Solution.

7. $y = \dfrac{1}{x + 2}$

$-7 \le x \le 3, -4 \le y \le 4$

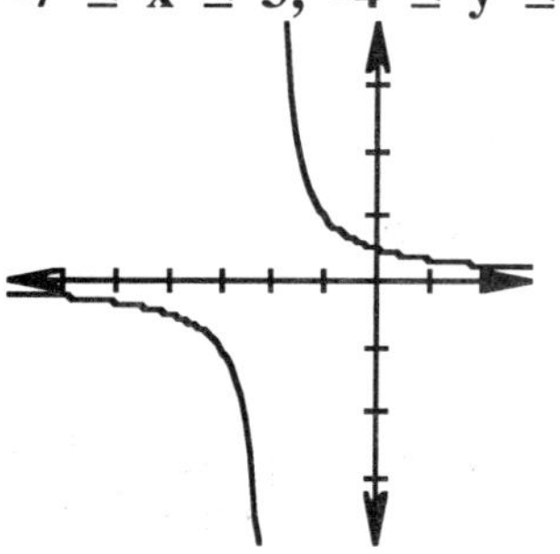

8. $y = \dfrac{x + 2}{x - 3}$

$-5 \le x \le 10, -6 \le y \le 6$

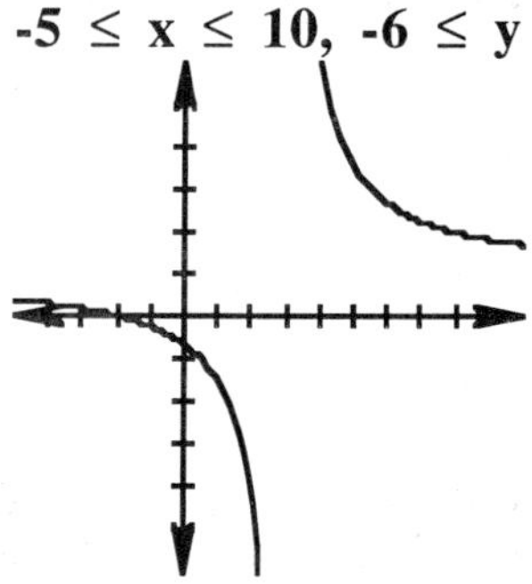

If you are not getting the same graph even with the same window indicated above, then chances are you didn't type the equation in correctly. For number 7, you must type y = 1/(x + 2), and you **must** have the parentheses! If you type y = 1/x + 2 your graph will have the same shape but it will not be in the same position. Similarly with number 8, you must type y = (x + 2)/(x - 3), and include **both** sets of parentheses. Entering y = x + 2/x - 3 will give you something completely different and incomplete.

Graphing and using your graphing calculator are necessary skills, but these are just stepping stones to the bigger problems: interpreting graphs. You need to understand what a graph represents, particularly within a real world situation. This is what you were studying in the Shrinking Arrows and Shrinking Targets labs. Let's look at another.

9. **A 50 gallon barrel is filled to the brim with water. The water depth is 38 inches. A small plug is at the bottom of the barrel. A student pulls the plug and lets the water run out. After three minutes, the depth is 32.75 inches. After seven minutes the depth is 25.75 inches. Graph the data. What does the slope of the data tell you? What does the x-intercept tell you?**

Solution.
Once the plug is pulled, the water level will drop at a steady rate until the barrel is empty. The three data points we pulled from the description, (0, 38), (3, 32.75), and (7, 25.75), are shown on the graph in figure 1. They appear to be linear and in fact, in figure 2, I drew in a "best fitting" line. The problem asks us about the slope and the x-intercept. I could approximate each of these from the graphs, and in some cases, that would be a suitable method of solving the problem. But, I will use the data points to find the equation of the line and use the equation to find the x-intercept.

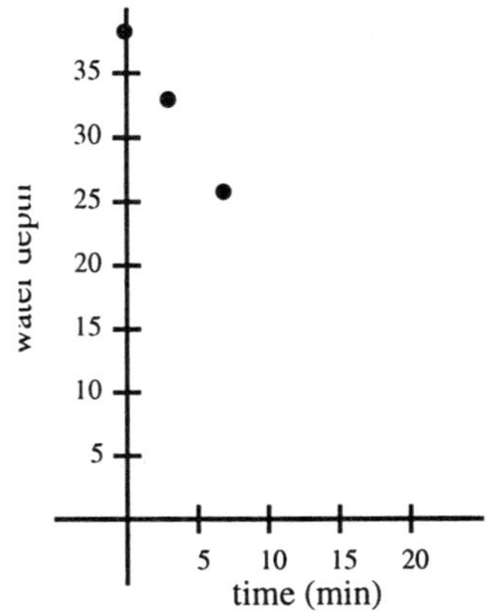

figure 1

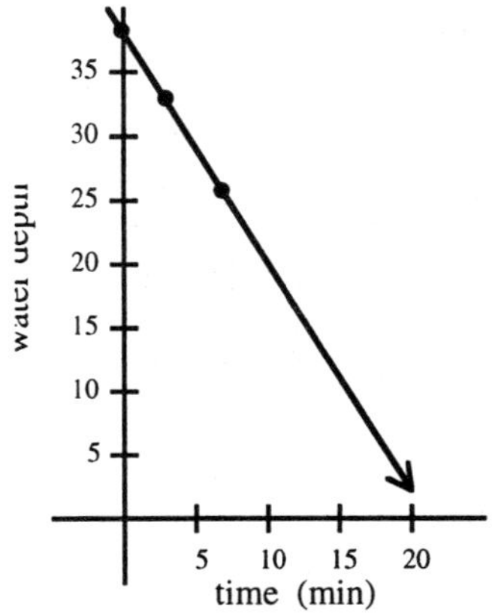

figure 2

Using the first two points, (0, 38) and (3, 32.75), I calculate the slope to be

$$m = \frac{32.75 - 38}{3 - 0} = -1.75.$$

Since we already know the y-intercept we can now write the equation:

$$y = -1.75x + 38.$$

Now, the problem: what does the slope tell you? Yes we know the slope is -1.75 but what does this number have to do with the problem? Is it a random number? No it is not. It is actually an important number in the problem. The key to discovering its meaning lies on the graphs. Notice how the axes are labeled: the vertical (y) axis is labeled "water depth" while the horizontal (x) axis is labeled "time." The slope has these same "units:" we took the change in y (the change in water depth) over the change in x (the change in time). So the units of the slope is $\frac{\text{water depth}}{\text{time}}$. In other words, the slope tells us how much the water drops each minute: 1.75 inches. The negative sign indicates that it is dropping.

To find the x-intercept we let y = 0 in our equation and solve for x. (This is something you must know how to do! If you can't remember it -- or even if you can -- put the first sentence of this paragraph in your tool kit along with an example!)

$$\begin{aligned} 0 &= -1.75x + 38 \\ -38 &= -1.75x \\ x &\approx 21.71 \end{aligned}$$

The x-intercept is approximately (21.71, 0). What does this number mean? For a hint, look at the units of the axes again: x is time in minutes and y is the water depth. At the x-intercept the y coordinate (the water depth) is zero. Therefore, this point tells us how long it will take for the water to run out: approximate 21.71 minutes.

A key idea in Algebra 2 is realizing that points on a graph <u>mean</u> something. You need to think about what a point means and make sense out of it. It will be a long time before a graphing calculator will be able to this so you are on your own here!

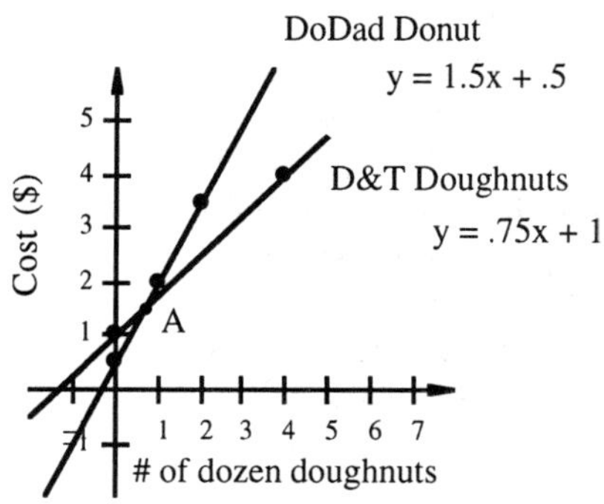

Consider the graph at right. We can (and will!) ask numerous questions about this graph and glean a great deal of information from it. First, a general question: **what does the graph tell us?** (Answer this on your own before you continue reading!) It tells us the cost of dozens of doughnuts at two different places. **What does point A represent?** This is a key question! Point A is a special point in that it is on both lines at the same time. Since point A is on both graphs, point A satisfies both equations. To find out the exact coordinates of point A, we use the two equations of the lines to solve for x and y.

$$\left.\begin{array}{l} y = 1.5x + .5 \\ y = .75x + 1 \end{array}\right\} \quad 1.5x + .5 = .75x + 1$$

We use substitution to write one equation and now we solve for x.

$$\begin{aligned} 1.5x + .5 &= .75x + 1 \\ .75x + .5 &= 1 \\ .75x &= .5 \\ x &= 0.666666\ldots \\ x &\approx 0.67 \end{aligned}$$

The x value is $\frac{2}{3}$ -- it probably would have been easier to solve this with the numbers as fractions rather than decimals! Of course, we are not done! Now we must solve for y. This time I will use the fraction. Remember, with substitution, I can substitute the x value back into either equation. Try to choose the one that will make things easier for you. (Sometimes there is no clear cut winner here!)

$$\begin{aligned} y &= \frac{3}{4}x + 1 \\ y &= \frac{3}{4}\left(\frac{2}{3}\right) + 1 \\ y &= \frac{1}{2} + 1 \\ y &= \frac{3}{2} \text{ or } 1.5 \end{aligned}$$

So point A has coordinates $\left(\frac{2}{3}, \frac{3}{2}\right)$. What does this mean? Remember the units of the axes and that point A is on both graphs. This point tells us that two-thirds of a dozen doughnuts cost \$1.50 at <u>both</u> places.

Another question you should be able to answer: **What do the slope and y-intercept of each line tell you?** The units of the axes are the key to deciphering the meaning of the slope. These units are $\frac{\text{cost}}{\text{\# dozen}}$ or in more common terms, the cost per dozen doughnuts. Therefore, at DoDad Donuts, the cost per dozen doughnuts is $1.50 while at D&T Doughnuts the cost is only $.75. To figure out the meaning of the y-intercept, recall what the coordinate of the y-intercept is. For instance, for D&T Doughnuts, the y-intercept is the point (0, 1). Since the x coordinate is zero, this point is saying that zero doughnuts cost $1.00. This represents a cost the shop has that is not the cost in preparing doughnuts. One possibility is that this is the cost of the box or other packaging for the doughnuts. At DoDad Donuts, the cost of its packaging is just $.50.

In all problems, but in particular, problems in which you are asked to explain or justify, show all your work and explain clearly and completely. Let the teacher know everything you know. This is very important in problems that ask you to investigate a function. In those problems, your teacher is looking for what you know, not what you don't know. You must include every detail you can. Yes, your teacher knows the math and might be able to figure out what you are going to say or what you might have left out, but assume that's not the case. A rule of thumb: write your solutions and explanations as if you are writing it to a parent or a sibling who is not in the class. Be clear. It helps the teacher give you the appropriate grade, but more importantly, it will strengthen your understanding.

UNIT 1
MORE TO TRY

Solve the following for x and/or y.

1. $5(x + 7) = -2x - 10$
2. $6 - (x + 3) = 9 - 3x$
3. $3.1x - 7 = -.6x + 4$
4. $3x + y = 12$; $y = 3x$
5. $x - 6y = 10$; $y = \frac{1}{3}x + 2$
6. $7 - 3x = 15 + 2x$
7. $ax + b = c$
8. $b(x - a) = c$
9. $3x^2 - 6x - 1 = 0$
10. $x^2 - 4x = 21$
11. $12x^2 - 35x + 8 = 0$
12. $x^2 - 32 = 0$

Find the error and show the correct solution.

13.
$$\begin{aligned} 5x - 9 &= -2(x - 3) \\ 5x - 9 &= -2x + 5 \\ 7x &= 14 \\ x &= 2 \end{aligned}$$

14.
$$\begin{aligned} 8x^2 + 4x &= 12 \\ 2x^2 + x &= 3 \\ 2x(x + 1) &= 3 \\ 2x = 3 \quad &\text{or} \quad x + 1 = 3 \\ x = \tfrac{3}{2} \quad &\quad\quad x = 2 \end{aligned}$$

Sketch a complete graph of each of the following. Be sure to label the graph carefully enough so that all key points are identified.

15. $y = -.1x + 30$ 16. $y = -x^2 + 30x - 445$ 17. $y = \frac{3x}{x - 15}$

18. If $f(x) = 3x^2 - 6x$, find
a) $f(1)$ b) $f(-3)$ c) $f(2.75)$

19. If $g(x) = -.3x + 6.3x^2$, find
a) $g(-2)$ b) $g(.4)$ c) $g(18)$

20. If $t(n) = \frac{2n}{n + 1}$, find
a) $t(-4)$ b) $t(9)$ c) $t(-1)$

Use the graph below right to answer the following questions.

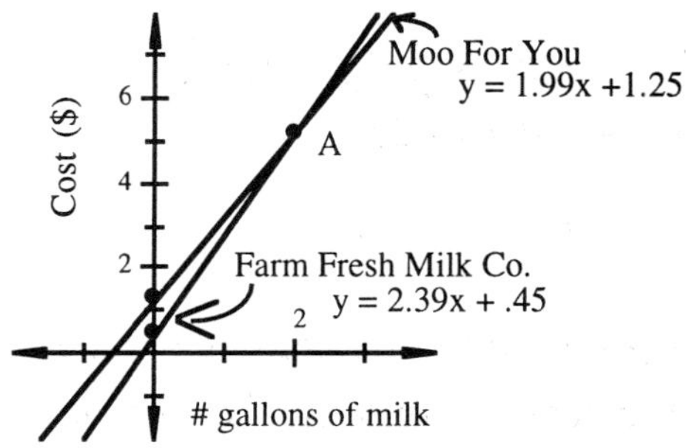

21. Briefly, what does this graph represent?

22. Explain the significance of point A. What is the coordinate of point A?

23. For each equation, explain the meaning of the slope.

24. For each equation, explain the meaning of the y-intercept.

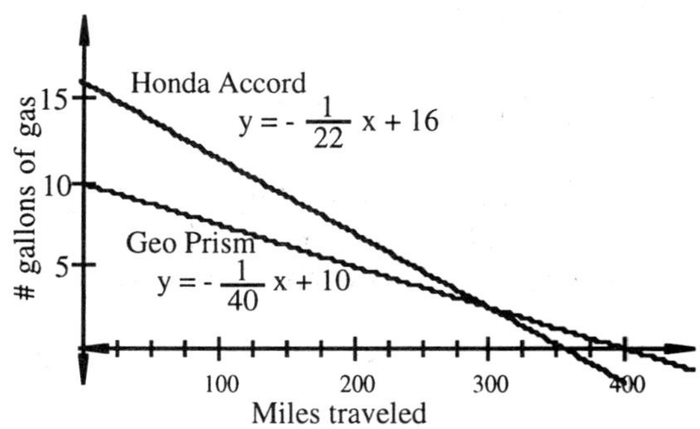

25. Briefly, what does this graph represent?

26. Based on the graph, give all the information you can about the Honda Accord.

27. Based on the graph, give all the information you can about the Geo Prism.

28. Does it make sense to extend these lines into the second and fourth quadrants? Explain.

29. Investigate the function $y = \sqrt{x - 6} + 1$.

30. Investigate the function $y = \frac{1}{x - 6}$.

SAT PREP

These problems are very similar to actual SAT test questions. Use a calculator whenever you need to. On multiple choice questions, choose the best answer from the ones provided. When a picture has a diagram, assume the diagram is drawn accurately except when a problem says it is not. These questions do not just cover the material you have done in class so far, so don't be surprised if these look different. Good luck!

1. If $x + 9$ is an even integer, then which of the following could be the value of x?

   a) 4 b) 2 c) 0 d) -1 e) -2

2. If $(m + 5)(11 - 7) = 24$, then m =

   a) 1 b) 4 c) 8 d) 11 e) 17

3. The fractions $\frac{3}{d}$, $\frac{4}{d}$, and $\frac{5}{d}$ are in simplest reduced form. Which of the following could be the value of d?

   a) 20 b) 21 c) 22 d) 23 e) 24

4. A group of three numbers is called a "j-triple" for some number j, if $\left(\frac{3}{4}j\,,\ j\,,\ \frac{5}{4}j\right)$. Which of the following is a j-triple?

   a) (0, 4, 5) b) $\left(5\frac{3}{4}, 6, 6\frac{1}{4}\right)$ c) (9, 12, 15)

   d) (750, 1000, 1250) e) (575, 600, 625)

5. A ball is thrown straight up. The height of the ball can be modeled with the equation $h = 38t - 16t^2$, where h is the height in feet and t is the number of second since the ball was thrown. How high is the ball two seconds after it is thrown?

   a) 12 b) 16 c) 22 d) 32 e) 40

6. In the figure at right, AC is a line segment with a length of 4 units. What is the value of k?

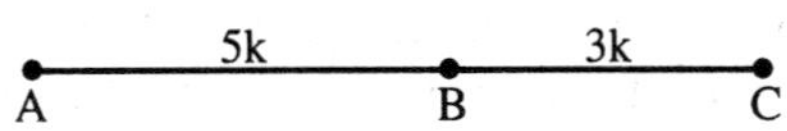

7. Let the operation § be defined as a § b is the sum of all integers between a and b. For example, 4 § 10 = 5 + 6 + 7 + 8 + 9 = 35. What is the value of (130 § 170) - (131 § 169)?

8. An isosceles triangle has a base of length 15. The length of each the other two equal sides is an integer. What is the shortest possible length of these other two sides?

9. Assume that $\frac{1}{4}$ quart of cranberry concentrate is mixed with $1\frac{3}{4}$ quarts of apple juice to make cranapple juice for four people. How many quarts of cranberry concentrate are needed to make a crabapple drink at the same strength for 15 people?

10. A stack of five cards is labeled with a different integer ranging from 0 to 4. If two cards are selected at random without replacement, what is the probability that the sum will be 2?

UNIT 2

The Bouncing Ball
SEQUENCES

UNIT 2
THE BOUNCING BALL AND OTHER SEQUENCES

There is more review in this unit although the topic, sequences, is new. You will find that while you are learning about sequences the review will be embedded within the new problems. Just remember: think similar, simpler, problems!

A necessary skill you must have to work efficiently with sequences is simplifying exponents. You should have the Laws of Exponents, with examples, in your tool kit. Consider the following examples. (Remember to try them first!)

1. Simplify the following expressions.

a) $(4x^3)^5$ b) $(7x^2)(-6x^4)$ c) $\frac{6x^3y^5}{2x^2y^3}$ d) $\frac{4(x^3y)^5}{(2xy)^2}$

Solution.

To begin with, let's review the Laws of Exponents:

1. $x^ax^b = x^{a+b}$ 2. $\frac{x^a}{x^b} = x^{a-b}$ 3. $(x^a)^b = x^{ab}$ 4. $(xy)^a = x^ay^a$

Although these four laws might look much simpler than the problems above, this is all you need to simplify parts (a) through (d). I will state which Law I used at key steps.

a) $(4x^3)^5$
$= (4)^5(x^3)^5$ by Law 4
$= 1024x^{15}$ by Law 3

b) $(7x^2)(-6x^4)$
$= -42x^2x^4$
$= -42x^6$ by Law 1

c) $\frac{6x^3y^5}{2x^2y^3}$
$= 3x^{3-1}y^{5-3}$ by Law 2
$= 3x^2y^2$

d) $\frac{4(x^3y)^5}{(2xy)^2}$
$= \frac{4x^{15}y^5}{2^2x^2y^2}$ by Law 4
$= x^{15-2}y^{5-2}$ by Law 2
$= x^{13}y^3$

Besides knowing the Laws of Exponents, you should also be able to justify why they are true. For instance, the first law, $x^ax^b = x^{a+b}$, can be verified this way:

$$x^ax^b = \underbrace{(x \cdot x \cdot x \cdot \ldots \cdot x)}_{\text{a of these}}\underbrace{(x \cdot x \cdot x \cdot \ldots \cdot x)}_{\text{b of these}} = \underbrace{x \cdot x \cdot x \cdot \ldots \cdot x}_{\text{a + b of these}} = x^{a+b}$$

This justification is based on the definition of an exponent that $x^a = x \cdot x \cdot x \cdot \ldots \cdot x$, a times. The third law is explained this way:

$$(x^a)^b = \underbrace{x^a \cdot x^a \cdot x^a \cdot \ldots \cdot x^a}_{\text{b of these}} = x^{a+a+a+\ldots+a} = x^{ab}$$

The next review topic is solving two equations with two unknowns. You might be solving these problems by graphing or by guess and check. Although these methods give you lots of information, they are not always the best methods. The can be inefficient and inaccurate. In the last section of this guide we briefly reviewed the substitution method of solving two equations with two unknowns. Now we will look at the elimination method. Consider this system.

3. Solve for a and d.

$$\begin{aligned} -3 &= 5d + a \\ 357 &= 50d + a \end{aligned}$$

Solution.

First, don't panic because this problem has "a"s and "d"s, as you will see them more often in this unit. If it really bothers you, rewrite the equations with "x"s and "y"s: $-3 = 5x + y$ and $357 = 50x + y$.

We will use the elimination method to solve this system. In the elimination method, we add or subtract the equations, or a multiple of the equations, to <u>eliminate</u> one of the variables. Let the name of the method help you remember the procedure! If nothing disappears when you are using this method, you need to check your work or modify the equations. Watch what happens when I subtract the bottom equation from the top.

$$\begin{aligned} -3 &= 5d + a \\ \underline{357} &\underline{= 50d + a} \\ -360 &= -45d \end{aligned}$$

Good! The "a" is eliminated. We are on the right track. Now all that remains is a single variable equation. You've solved equations like this for years! Dividing both sides by -45,

$$d = 8$$

Of course, we are not done. Remember these are **two** equations with **two** unknowns. We have another variable to find. We substitute the value of d back into either equation. I'll use the top one -- it looks easier since the numbers are smaller.

$$\begin{aligned} -3 &= 5(8) + a \\ -3 &= 40 + a \\ a &= -43 \end{aligned}$$

Therefore the solution is $d = 8$ and $a = -43$. And what we have actually done is answered BB-41. When you read the problem you might have thought it was strictly a sequence problem, but that was just the "outer layer". Digging deeper reveals a type of problem you already know how to solve. Thus, review is embedded within the new topic!

Here is a similar problem.

4. A line goes through the points (-3, 10) and (5, 2). What is the equation of the line?

Solution.
The equation of a line looks like: $y = mx + b$. We have two points that give us values for x and y. Using the first point we can write:

$$10 = -3m + b$$

And using the second point we write

$$2 = 5m + b$$

Now we solve these two equations with two unknowns, m and b, to find the equation of the line. Subtracting again, (you are remembering to try these yourself, right?)

$$\begin{aligned} 10 &= -3m + b \\ \underline{\ \ 2} &\underline{= 5m + b} \\ 8 &= -8m \\ m &= -1 \end{aligned}$$

Substituting back into the top equation (remember either one will work),

$$\begin{aligned} 10 &= -3(-1) + b \\ 10 &= 3 + b \\ b &= 7 \end{aligned}$$

Since $m = -1$ and $b = 7$, the equation of the line is $y = -1x + 7$. How can we check that this is correct? Substitute each of the two original points into this equation and make sure it makes the equation true.

(-3, 10)

$$\begin{aligned} 10 &\stackrel{?}{=} -1(-3) + 7 \\ 10 &= 3 + 7 \quad \surd \end{aligned}$$

(5, 2)

$$\begin{aligned} 2 &\stackrel{?}{=} -1(5) + 7 \\ 2 &= -5 + 7 \quad \surd \end{aligned}$$

Note: If you are going to check (and you <u>are</u> going to check!), you must check the equation with <u>both</u> points!

A key part of learning algebra is about <u>making the connection</u>. Algebra has meaning. It is about seeing how different components connect to help you solve problems. Algebra (although at times you may not believe this!) should make problems <u>easier</u>. If you can see how problems are related, sometimes even the same, you are understanding the mathematics.

Let's look at another problem, one in which algebraic notation makes the problem easier to write.

5. Consider the sequence: $t(n) = 8 + 7n$.

What are t(0), t(1), t(2), and t(3)?

Solution.
This notation, function notation, t(n), is an algebraic representation of the sequence. It lets us write the sequence compactly. To find t(0) if t(n) = 8 + 7n, simply replace n with 0:

$$t(0) = 8 + 7(0)$$
$$t(0) = 8$$

That's all you do with function notation!

$$t(1) = 8 + 7(1) = 15$$
$$t(2) = 8 + 7(2) = 22$$
$$t(3) = 8 + 7(3) = 29$$

We could even ask for t(♥), t(♥) = 8 + 7(♥)

or t(q), $t(q) = 8 + 7(q)$

Of course we can't simplify these further, but we do have an answer. Once substituted, be sure you simplify correctly when it is appropriate.

6. For the same sequence described in problem 5, is 2371 one of its terms? Explain how you know.

Solution.
How could we find out whether or not 2371 is one of the terms in this sequence? You probably wouldn't want to list all the terms of the sequence (8, 15, 22, 29, 36, ...) until you got to 2371. That's a lot of terms! And if you did, could you be sure you didn't make any errors in all of those "plus 7"s? Let's use algebra to make this problem easier, and to be sure of our answer. Asking whether or not 2371 is a term is the same as asking if there is an "n" such that

$$8 + 7n = 2371$$

Let's solve for n and find out!

$$7n = 2363$$
$$n = 337\frac{4}{7}$$

We solved for n and got this (not so pretty) answer. Does this mean 2371 is a term in the sequence? NO! Remember, this is a sequence. The variable n represents the **term number**: t(0) is the initial value, t(1) is the first term after the initial value, t(2) is the second term after the initial value, etc. Is there a $337\frac{4}{7}$th term? No, the domain of these functions, sequences, is the whole numbers: 0, 1, 2, 3, ... So 2371 is not a term in this sequence. Using algebra is the most efficient and trustworthy way to show this.

The previous examples have illustrated arithmetic sequences, sequences in which you add the same number (called the common difference) each time to get the next term. There is another type of sequence called geometric. You should know the difference between the two, know how decide which type is which, and know the characteristics of each. And this should all be in your tool kit! Be sure to try the next two problems before looking at the solution.

7. How are arithmetic and geometric sequences similar?

Solution.
Both of these sequences are **discrete** functions. (Do you know what discrete means? This means their graphs are not connected like a straight line, but are just points. A straight line or a curve that is connected is called continuous.) The domain of both types is the whole numbers (0, 1, 2, 3, ...) and the range for both is the sequence itself. Both can be represented with an equation or recursively. Both can be represented on a graph.

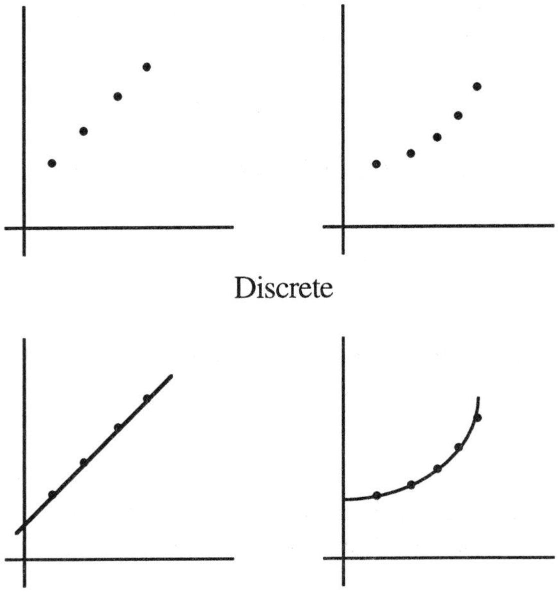

8. How are geometric and arithmetic sequences different?

Solution.
The graph of an arithmetic sequence is linear (the points may not be connected, but they lie in a straight line) while the points of a geometric sequence lie on a curve. The terms of an **arithmetic sequence** are formed by **adding** the same number (called the common difference) to each term to get the next, while the terms of a **geometric sequence** are found by **multiplying** the same number (known as the common ratio) to each term.

These terms, **common difference** and **common ratio**, should be in your tool kit, but they should also make sense to you. They should help you remember the key difference between the two types of sequences and help you identify these types. Consider this next problem.

9. Identify each sequence as arithmetic, geometric, both or neither.

a) t(n) = 3, 6, 12, 24, ... **b) s(n) = 8, 20, 32, 44, ...**

c) r(n) = 1, 2, 5, 10, ... **d) u(n) = 1, 1, 1, 1, ...**

Solution.

a) t(n) = 3, 6, 12, 24, ...

To determine what type of sequence each of these is, start by looking for a common difference and then a common ratio. To find a common difference, do just as the name implies: take the difference between consecutive terms and see if they have the difference in common.

$$6 - 3 = 3, \quad 12 - 6 = 6$$

Not a common difference. Next take ratios.

$$\frac{6}{3} = 2, \quad \frac{12}{6} = 2$$

Aha! Let's check one more to be sure.

$$\frac{24}{12} = 2$$

Since these terms have a common ratio, this sequence is geometric. The common ratio is 2.

b) s(n) = 8, 20, 32, 44, ...

Again, start by looking for a common difference.

$$20 - 8 = 12, \quad 32 - 20 = 12$$

Good! Check one more to be sure.

$$44 - 32 = 12$$

So this is arithmetic. Could it also be geometric? Check the ratios.

$$\frac{20}{8} = 2.5, \quad \frac{32}{20} = 1.6$$

So this sequence is just arithmetic. The common difference is 12.

c) r(n) = 1, 2, 5, 10, ...

Start by checking the common difference.

$$2 - 1 = 1, \quad 5 - 2 = 3$$

Not arithmetic! What about geometric? Check the ratios.

$$\frac{2}{1} = 2, \quad \frac{5}{2} = 2.5$$

This sequence is neither arithmetic nor geometric, and there is no common difference nor common ratio.

d) u(n) = 1, 1, 1, 1, ...

Start by checking the common difference.

$$1 - 1 = 0, \quad 1 - 1 = 0$$

This is arithmetic! Now check the ratios.

$$\frac{1}{1} = 1, \quad \frac{1}{1} = 1$$

And this is geometric too! In fact, any sequence that is just a constant term is both arithmetic and geometric. This sequence is arithmetic with common difference 0 and geometric with common ratio 1.

Make sure that you check for both arithmetic and geometric because as you can see, a sequence can be both or neither.

Now let's look at two typical applications of sequences.

10. On the day Tess was born, her dad put a five gallon water jug in her room and dropped 22 quarters into it. On each birthday since then, he increased the number of quarters he dropped in by 17, so on her first birthday, he dropped 39 quarters in, on her second he added 56 quarters.

birthday	# quarters
0	22
1	39
2	56
3	
4	
5	
6	

a) Complete the table showing the number of quarters dropped in each birthday.

b) Is the sequence formed by the number of quarters arithmetic, geometric, both or neither? Explain how you know.

c) Write an equation representing this sequence. Graph this sequence.

d) How much money is in the jug when Tess turns 15?

e) Will Tess have $75 in her jug before she leaves home just after her 18th birthday? Will she ever have exactly $75 in the jug? Justify your answer.

Solution.

birthday	# quarters
0	22
1	39
2	56
3	73
4	90
5	107
6	124

a) Once you figure out how to fill in the table, you probably have answered part (b) as well. The number of quarters increases by 17 each time.

b) The sequence is arithmetic with common difference 17 since we add 17 each time. Let's just check to see if it is geometric as well. (Yes, we all know that it isn't but it is good practice to check!) Check for a common ratio: $\frac{39}{22} \approx 1.77, \frac{56}{39} \approx 1.43$. Not the same so this is only arithmetic.

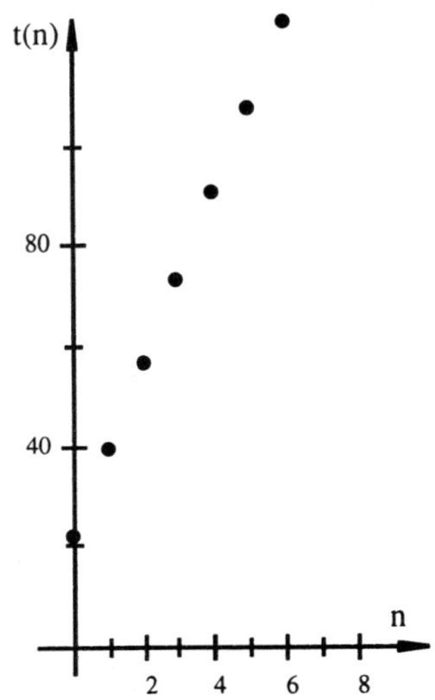

c) There are several ways to write the equation of this sequence. One way is as we did in problem 3 where we solved for a and d. Simply use any two points, (0, 22) and (1,39) will work nicely, and the equation

$$t(n) = dn + a.$$

Substitute the values for n and t(n) to write two equations with two unknowns.

$$22 = d(0) + a$$
$$39 = d(1) + a$$

This is particularly easy to solve since the first equation reduces to a = 22. Substituting this into the second equation gives d = 17. Therefore our equation is

$$t(n) = 17n + 22$$

Notice that the pattern for the equation for an arithmetic sequence is always

$$t(n) = dn + a$$

where d is the common difference and a is the initial value. So we could have written the equation down without having solved the system of equations. Of course, solving them was good practice!

d) To find out how much will be in there when Tess turns 15 we use the equation we just found and find t(15).

$$t(15) = 17(15) + 22$$
$$t(15) = 277$$

This tell us that when Tess turns 15 there will be 277 **quarters** in the jar. But how much money is that? Every four quarters is a dollar, so $\frac{277}{4} = 69.25$. Tess will have \$69.25 in her jar when she is 15.

e) As we just did, let's find t(18) to see how much she has when she is 18.

$$t(18) = 17(18) + 22$$
$$t(18) = 328$$

At 18, she has 328 quarters which is $\frac{328}{4}$ = \$82.00, so she does have \$75. Does she ever have exactly \$75? In other words, is there an n such that

$$300 = 17n + 22?$$

Solve for n to see.

$$278 = 17n$$
$$n \approx 16.35$$

Since the domain for n is only whole numbers, and 16.35 is not a whole number, Tess will never have exactly \$75. Do you know why the \$75 was written as 300 in the equation? Don't forget that the equation gives the number of quarters. \$75 is 300 quarters.

Let's do another application.

11. Davis and Emma are best friends. Upon graduating high school, Emma goes to work for her father making $28,000 per year. She gets a 1.25% raise each year. Davis doesn't start working until eight years later after finishing college and graduate school. He starts teaching at a small college making $35,000 per year, with annual raises of 1.75%. How much is each making fifteen years graduating high school?

Solution.
If you are thinking that this doesn't look like a problem about sequences, think again. This is a typical application problem on geometric sequences. Here's why: when they are working, each year they each get a raise. For Emma, she gets 1.25% each year. So after one year (as she is starting her second year) she is making 28,000(1.0125). (This is something you should know: we could calculate her percent increase by taking .0125 times her salary and then adding this to her salary, 28000 + .0125(28000), but this is the same as taking 1.0125 times the salary. See BB-50 and BB-70 and their solutions to see why this simplifies this way.) After two years, we multiply this new salary by 1.0125 again: $28000(1.0125)(1.0125) = 28000(1.0125)^2$. So each time we multiply the term by the same number -- that's the definition of a geometric sequence! After three years, we multiply by 1.0125 again: $28000(1.0125)^2(1.0125) = 28000(1.0125)^3$. Are you starting to see the pattern? Write out a few more to convince yourself that after fifteen years with her father's company, Emma will be making $28000(1.0125)^{15} \approx \$33735.22$.

Although Davis has eight years of no income, once he starts working, he makes $35,000 the first year, more than Emma makes after fifteen years of work. After one year of work (which is nine years after graduating high school), Davis makes 35000(1.1075). After two years (as he begins his third year and ten years after graduating high school), Davis makes $35000(1.0175)^2$. To find out how much Davis makes fifteen years after graduation we calculate $35000(1.0175)^7 \approx \$39519.28$.

Make sure that your tool kit is up to date with any new information you have gained this unit. Don't forget to include any new information you might have to add to your list for investigating a function. Sequences are functions, discrete functions, so be sure to include information about them in the list. By the end of this unit I think you will find that your algebra skills, particularly those of solving equations and system of equations, have improved. That's good! Those skills will be put to good use!

UNIT 2
MORE TO TRY

Simplify the following.

1. $7(x^3y^7)^2$
2. $(3x^5)(4x^2)^8$
3. $\dfrac{(6a^2b^9)^4}{2a^7b^{35}}$
4. $(5a^3x^7b)^6$
5. $\dfrac{-3(xy)^6z^{10}}{2x^5y^4z}$
6. $4\left[\dfrac{-5b^4}{a^3}\right]^3$
7. $\dfrac{(-2x^3y^2)^5}{-8x^7y^3}$
8. $-2(a^4x^5)(3a^6x^3)^5$

Each of the functions listed below defines a sequence. List the first five terms of the sequence and state whether the sequence is arithmetic, geometric, both or neither. Justify your response.

9. $t(n) = 5n + 2$ 10. $s(n) = 3 - 8n$ 11. $u(n) = 9n - n^2$ 12. $t(n) = (-4)^n$

13. $s(n) = \left(\frac{1}{4}\right)^n$ 14. $u(n) = n(n + 1)$ 15. $t(n) = 8$ 16. $s(n) = \frac{3}{4}n + 1$

Identify each of the following sequences as arithmetic or geometric. Then write the equation which gives the terms of the sequence.

17. 48, 24, 12, 6, 3, ... 18. -4, 3, 10, 17, 24, ... 19. 43, 39, 35, 31, 27, ...

20. $\frac{2}{3}, \frac{1}{2}, \frac{3}{8}, \frac{9}{32}, \frac{27}{128}, \ldots$ 21. 5, -5, 5, -5, 5, ... 22. 10, 1, 0.1, 0.01, 0.001, ...

Graph the following two sequences on the same set of axes.

23. $t(n) = -6n + 20$ 24. 1, 4, 16, 64, ...

25. Do the two sequences of the last two problems have any terms in common? Explain how you know.

Examine each graph below. State what type of sequence each represents, then find the equation of the sequence.

25.

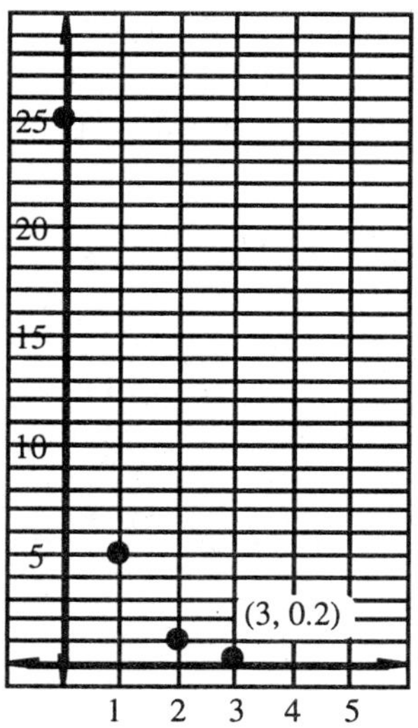

26.

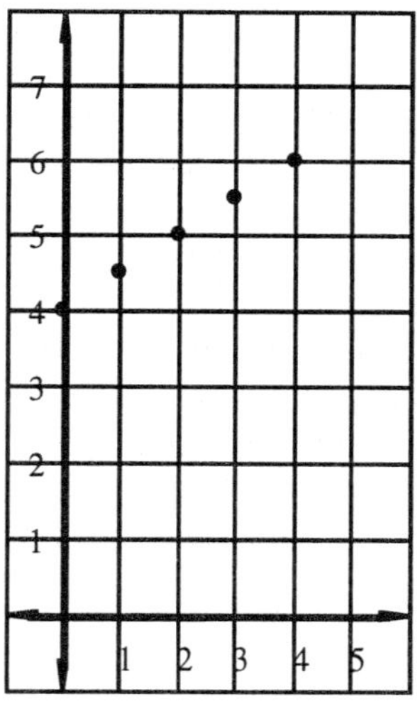

28. Every year, the average height for a man is 100.05% of what it was the previous year. In the year 1548, the average man's height was 54 inches. Based on this information, what is the average height for a man in 1998?

29. Davis has \$5.40 in his bank on his fourth birthday. If he adds \$.40 to his bank every week, when will he have enough to buy the new Smokin' Derby race car set which retails for \$24.99?

30. Investigate the sequence $t(n) = -4n + 18$.

SAT PREP

1. In the correctly worked addition problem at right, each Π represents the same digit. What is the value of Π?

$$\begin{array}{r} \Pi 6 \\ \Pi 7 \\ \Pi 8 \\ +\ \Pi 9 \\ \hline 190 \end{array}$$

a) 3 b) 4 c) 6

d) 9 e) 10

2. The figure at right is not drawn to scale. The cone has a height of 22 inches and a radius of 12 inches. The cone is cut parallel to the circular base creating two shapes, one of which is a smaller cone. If the radius of the base of the small cone is 4 inches, what is the height of the small cone?

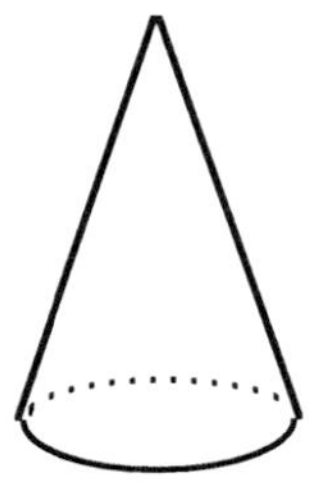

a) $6\frac{1}{2}$ b) 7 c) $7\frac{1}{3}$

d) 8.05 e) 8.99

3. The table at right shows the distances required to stop a car traveling at a given speed. The car will travel a certain distance while the driver is "thinking" about what to do. Once the driver applies the brake, the car travels some more before coming to a stop. How many more feet does it take to stop a car traveling at 60 miles per hour than at 20 miles per hour?

Speed (in miles per hour)	Thinking Distance (in feet)	Braking Distance (in feet)
20	20	20
30	30	45
40	40	80
50	50	125
60	60	180

a) 75 b) 105 c) 150

d) 175 e) 200

4. Eight students played a game and their scores from least to greatest are:

2 , 3 , 5 , 6 , 7 , 8 , 8 , 9

Which of the following are true statements?

I. The average (arithmetic mean) is less than 7.
II. The median is less than 7.
III. The mode is greater than 7.

a) None b) III only c) I and II only

d) II and III only e) I, II, and III

5. The graph at right represents data taken at a local company. Which of the following is the closest approximation to the percent of employees of this company who travel for at least 16 minutes to get to work?

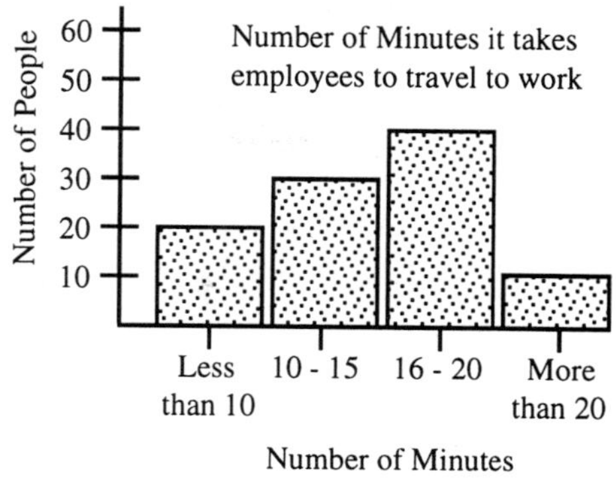

a) 25%

b) 30%

c) 40%

d) 50%

e) 60%

For the next five problems, compare the items in Column A and Column B. Answer

- A if the item in Column A is greater;
- B if the item in Column B is greater;
- C if the two quantities are equal;
- D if you cannot not tell from the information given.

Note: information pertaining to <u>both</u> columns is given in the middle.

	Column A		Column B
6.	$\frac{3}{2} - \frac{1}{2}$		$\frac{7}{8} - \frac{1}{8}$
7.	a	$ab = 0$	0
8.	x^6	$x^3 = y$	y^2
9.	o	The average (arithmetic mean) of m and n is greater than the average of m and o.	n
10.	r	$q + r = 76$ $q \neq 5$	71

UNIT 3

Fast Cars and Depreciation

EXPONENTIAL FUNCTIONS

UNIT 3
EXPONENTIAL FUNCTIONS

There is lots in this unit, but if you look for the connections (between these topics and previous topics) understanding the math will be easier. Just as we saw the connection between arithmetic sequences and linear functions last unit, we will see the connection between geometric sequences and **exponential functions** this unit. The applications this unit, growth and decay, are important, but don't forget that these are just sequence problems recast in a different setting. Some terms are changed: common ratio becomes the multiplier, initial value is called the principle in problems involving money and interest.

Let's start off with two typical examples.

1. Tess has \$500 with which to open a savings account. She can open an account at Fredrico's Bank which pays 7% interest, compounded monthly, or Money First Bank which pays $7\frac{1}{4}$ % compounded quarterly. Tess plans to leave the money in the account, untouched, for ten years. In which account should she place the money? Justify your answer.

Solution.
The quick answer is that she should put the money in the account which will pay her the most interest over the ten years. Now the question becomes, which bank is that? Let's look at Fredrico's Bank first. The principle (initial value) is \$500. Since the bank pays 7% compounded monthly, that means the interest is $\frac{.07}{12}$ each month, or approximately .00583 in interest each month. Now, remember all those problems last unit and this unit asking you to simplify expressions such as x + .08x? (This one becomes 1.08x.) In interest problems, the principle, call it x, plus its interest, x + .00583x in this case, simplifies to 1.00583x. Therefore, if Tess' money is put into Fredrico's Bank, after one month she will have

$$500(1.00583) \approx \$502.92.$$

To calculate the amount at the end of the second month, we must multiply by 1.00583 again, making the amount

$$500(1.00583)^2 \approx \$505.85.$$

At the end of three months, the balance is

$$500(1.00583)^3 \approx \$508.80.$$

This will happen every month for ten years, which is 120 months. At the end of 120 months, the balance will be

$$500(1.00583)^{120} \approx \$1004.43.$$

Now we must do the similar calculation for Money First Bank. Its interest rate is higher, $7\frac{1}{4}$%, but it is only compounded and calculated quarterly, (quarterly means four times each year, or every three months) but it is calculated the same way. Hence, every quarter the bank calculates $\frac{.0725}{4} = .018125$ interest. At the end of the first quarter, Tess would have

$$500(1.018125) \approx \$509.06.$$

At the end of ten years, that's 40 quarters, Tess would have

$$500(1.018125)^{40} \approx \$1025.69.$$

Since Money First would pay her approximately $21 more in interest that Fredrico's Bank, she should put her money in Money First Bank.

Let's look at another problem using exponential functions.

2. Most homes appreciate in value, at varying rates, depending on the home's location, size and other factors. But if the home is used as a rental, it may actually depreciate. Suppose a house originally costing $150,000 depreciates at a rate of 8% per year.

a) What is the multiplier which will give the value of the house after one year? What is that value?

b) What is the value after ten years?

c) What is the "half-life" of the value of the house (that is, when has the house lost half of its value)?

Solution.

a) Rather than adding an interest amount, with depreciation, we are taking some of the value away. After one year, the value of the house is 150000 - .08(150000) which is the same as 150000(.92). Therefore the multiplier is .92 and the value after one year is $138000.

b) After ten years the value of the house is

$$150000(.92)^{10} = \$65158.27$$

c) When will the value drop to $75000? In ten years the value is less than that so it happens in less than 10 years. Let's record some values in a table to determine the answer.

# Years	House value
1	1380002
2	126960
3	116803.20
4	107458.94
5	98862.23
6	90953.25
7	83676.99
8	76982.83
9	70824.20

Shortly after the eighth year the house will be half its value. Note: if you try to write an equation and solve, you will have to use guess and check to solve at this point. The equation, $75000 = 150000(.92)^x$, requires some new mathematics to solve. We will learn how to solve such equations exactly in Unit 6. If you wish to try to solve this by guess and check, the answer you are looking for is approximately 8.313.

Along with growth and decay problems, you will get more practice simplifying exponents and will learn some new information about them. In problems FX-28 through FX-30, you see three different explanations of what a zero exponent means. You might have seen other explanations as well. You will need to know that anything raised to the zeroeth power is one and you should also be able to give a convincing argument explaining why that is true. We also look at other exponents, negative and non-integer (fractions), and as with zero exponents, you should be able to justify why the rules work.

Briefly, here are the two additional laws you need to know about exponents. You should add them to your tool kit after the first four.

5. $x^0 = 1$ for all $x \neq 0$

6. $x^{-a} = \frac{1}{x^a}$

You shouldn't have any problems with number 5, but be careful with negative exponents. Many people make the mistake of thinking that a negative exponent makes a negative answer. Not so! The answer may be negative but not solely because of the exponent. As illustrated above and in FX-44, $2^{-3} = \frac{1}{2^3}$. A negative exponent calls for the reciprocal of the base. (Remember: the 2 in the last example is called the base.)

$$7^{-4} = \frac{1}{7^4}, \quad 10^{-2} = \frac{1}{10^2}, \quad 6^{-5} = \frac{1}{6^5}.$$

Notice that these results are **not** negative! The next question is, can the result ever be negative? Sure, when the base is negative **and** the exponent is odd. For example,

$$(-2)^{-5} = \frac{1}{(-2)^5} = \frac{1}{-32}$$

Do you see why the exponent must be odd? If it were even, a 4 instead of the 5, the answer is positive. (If you don't believe or understand this, convince yourself by multiplying (-4)(-4)(-4)(-4).)

What happens if the base becomes more complicated? If you remember the basic rules, you can always figure it out. For example, consider this next one.

$$\left(\frac{2}{3}\right)^{-3} = \frac{1}{\left(\frac{2}{3}\right)^3} = \frac{1}{\left(\frac{8}{27}\right)}$$

In case you have forgotten, the large bar in the fraction means "divide" so this messy fraction can be written as a division problem, the follow the rule “invert and multiply:

$$\frac{1}{\left(\frac{8}{27}\right)} = 1 \div \frac{8}{27} = 1 \cdot \frac{27}{8} = \frac{27}{8} \quad \text{therefore,} \quad \left(\frac{2}{3}\right)^{-3} = \frac{27}{8} = \frac{3^3}{2^3}$$

There's the reciprocal!

Now that zero and negative exponents have been explained, let's look at fractional exponents. Can we write $16^{1/2}$ and have it mean something? Yes, we can, and we will! The first thing to remember when dealing with fractional exponents is that they <u>are exponents</u>. So **don't** write $16^{1/2} = 8$. That's not true and it is a common error. Let's use the Laws of Exponents to help us understand what $16^{1/2}$ means. Recall that

$$x^a \cdot x^b = x^{a+b}.$$

So this means $$16^{1/2} \cdot 16^{1/2} = 16^1 = 16.$$

Also, recall that $$\sqrt{16} \cdot \sqrt{16} = 16.$$

For both of these to be true, it must also be true that $16^{1/2} = \sqrt{16}$, or $16^{1/2} = 4$.

Similarly $$16^{1/3} \cdot 16^{1/3} \cdot 16^{1/3} = 16^{1/3 + 1/3 + 1/3} = 16^1 = 16$$

This tells us that $16^{1/3} = \sqrt[3]{16}$ called "the cube root of 16."

To understand an exponent such as in $16^{2/3}$, recall another law of exponents: $(x^a)^b = x^{ab}$.

So $16^{2/3} = (16^{1/3})^2$ or $(16^2)^{1/3}$ since $\frac{1}{3} \cdot 2 = 2 \cdot \frac{1}{3}$. Also, $16^{2/3} = \sqrt[3]{16^2}$ or $\left(\sqrt[3]{16}\right)^2$.

There are several different ways to write these expressions and you will need to be familiar with all of them. (A typical test question is to have you rewrite an expression in as many different ways that you can.) Make sure that you can calculate such expressions on your calculator. For that, use the form with which you are most comfortable. A smart thing to do would be to list examples in your tool kit. All this will help you in solving equations such as in this next example.

3. Solve the following equations for x.

a) $8^{3x} = 2^{5x - 1}$

b) $27^{x + 5} = 9^{2x - 9}$

Solution.

To solve these you use a simple fact: if $a^x = a^y$, then $x = y$. This means we must try to make the bases the same on each side of the equal sign. This will take some thought on your part. Ask yourself what factor the bases have in common.

a) $8^{3x} = 2^{5x - 1}$

As suggested, let's ask the question: what factor do 2 and 8 have in common? 2. In fact, $8 = 2^3$. We will use this to rewrite the left side.

$$8^{3x} = (2^3)^{3x}$$

Now we can write

$$(2^3)^{3x} = 2^{5x - 1}$$

And by one of the laws of exponents, the left side can be rewritten once more.

$$2^{9x} = 2^{5x - 1}$$

Now we are set. The only way for these two expressions to be equal is for the exponents to be equal. Now we are back to familiar territory.

$$9x = 5x - 1$$
$$4x = -1$$
$$x = -\frac{1}{4}$$

b) $27^{x + 5} = 9^{2x - 9}$

Let's start with the question: what factor do 27 and 9 have in common? 3. Now, see if you can justify each step in this solution.

$$27^{x + 5} = 9^{2x - 9}$$
$$(3^3)^{x + 5} = (3^2)^{2x - 9}$$
$$3^{3x + 15} = 3^{4x - 18}$$
$$3x + 15 = 4x - 18$$
$$-x = -33$$
$$x = 33$$

Of course, just as with earlier problems, you should check your solutions to be sure they are correct.

Check:

$$8^{3(-1/4)} \stackrel{?}{=} 2^{5(-1/4) - 1}$$
$$8^{-3/4} \stackrel{?}{=} 2^{-9/4}$$
$$.210224 \ldots = .210224 \ldots \quad \surd$$

Check:

$$27^{33 + 5} \stackrel{?}{=} 9^{2(33) - 9}$$
$$27^{38} \stackrel{?}{=} 9^{57}$$
$$2.465 \ldots \cdot 10^{54} = 2.465 \ldots \cdot 10^{54} \quad \surd$$

It is important to see that these problems are different from the next examples. The problems you just did have the variable in the exponent. Notice where the variable is in the following problems.

4. Solve the following for x.

a) $x^7 = 42$ **b)** $3x^{12} = 132$

Solution.

Since the variable is the base and not the exponent, we don't have to get common bases as we did in the last examples. Now we just need to isolate the variable (get the variable by itself) and then get rid of the exponent. To get rid of the exponent we will actually use more exponents. Just remember how you would solve this equation:

$$x^2 = 329$$

You would solve this by taking the square root of both sides. As we found out earlier, the square root is the same as raising something to the $\frac{1}{2}$ power.

$$\left(x^2\right)^{1/2} = (329)^{1/2}$$

By one of our laws of exponents, the left side becomes

$$x^{2(1/2)} = \sqrt{329}$$

Since $2\left(\frac{1}{2}\right) = 1$, we have our solution: $x \approx \pm 18.138$

Watch how we use this same idea to solve these problems. Remember! Do the problems yourself first!

a) $x^7 = 42$

$$x^7 = 42$$
$$\left(x^7\right)^{1/7} = 42^{1/7}$$
$$x \approx 1.706$$

b) $3x^{12} = 132$

$$3x^{12} = 132$$
$$x^{12} = 44$$
$$\left(x^{12}\right)^{1/12} = 44^{1/12}$$
$$x \approx \pm 1.371$$

I'll leave it to you to check. Be sure you know how to operate your calculator efficiently. If you can, you will be able to check this type of problem quickly and accurately. One last point: notice how there is only one answer for part (a), where the exponent is odd, but there are two answers ($\pm$) in part (b) where the exponent is even. Even roots always produce two answers, a positive and a negative. Just be sure that if the problem is a real-world application that both make sense. You may have to throw out one to keep the answer feasible.

Next let's look at an application to review and lead to a new topic. Consider the following problem.

5. On a cold wintry day, as Karen made herself a cup of cocoa, she decided to conduct a mini science experiment as well. She took a mug and filled it with cocoa, placed a thermometer in the cocoa, and left it sitting next to her as she worked. Every couple of minutes or so, she wrote down the time and the reading on the thermometer. The data is shown below.

Time since 1st reading	**0**	**1**	**5**	**7**	**10**	**12**	**15**	**27**
Temp (°F)	**142°**	**128.7°**	**96.4°**	**88.01°**	**80.5°**	**77.6°**	**75.0°**	**72.2°**

a) Plot the points. Describe any trends you see.

b) What sort of function would best model/fit this data? Explain. What does the general equation of this sort of function look like?

Solution.

This particular graph seems to level off, and appears to become a horizontal line. In fact it doesn't become a horizontal line, but it does approach a **horizontal asymptote**. As you discovered with your team in FX-129 and FX-130, an asymptote is a line the function gets close to but never crosses. A function can have vertical and horizontal asymptotes. (Yes, more than one even!) In this example of cocoa cooling in a cup, does it make sense for the temperature to level off over time? It should! To what temperature would the cocoa level off? To the room temperature. We wouldn't expect a cup of cocoa -- without "outside help" -- to become cooler and cooler and cooler. (If so, who would need refrigerators?) From this data and from looking at the graph, we would conclude the room temperature to be approximately 72°. The best function to model or fit this data would be an exponential function. The general equation for an exponential function is $y = km^x$, if the function has a horizontal asymptote at $y = 0$. You will discover in Unit 6 that if the horizontal asymptote is elsewhere, the equation is $y = km^x + c$.

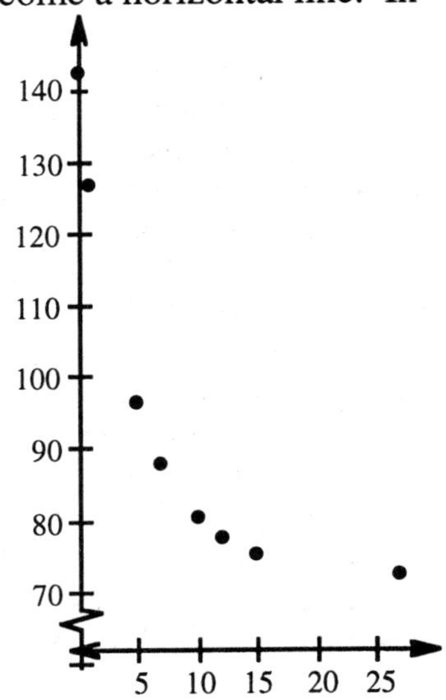

Once you have discussed asymptotes in your class, be sure to add all the important information to your tool kit. This is another piece of information to consider when you are investigating a function. See if you can find all of the asymptotes for the following function.

6. Investigate the following: $f(x) = \frac{1}{x^2 - 9}$

Solution.

If you use a graphing calculator to graph this, be sure you enter the equation correctly. You must type $1/(x^2 - 9)$. The parentheses are necessary. Your calculator might only confuse you initially, due to the many parts of this curve. So let's start gathering information. To find x-intercepts, we set y equal to zero and solve. Here,

$$0 = \frac{1}{x^2 - 9}$$

has no solution. (Note: the only way for a fraction to equal zero is for the numerator to equal zero. Since the numerator is fixed at 1, there is no way for it to equal zero. That's why the equation has no solution.) Therefore, there are no x-intercepts. What about y-intercepts? Set x = 0:

$$y = \frac{1}{0^2 - 9}, \text{ so } y = -\frac{1}{9}$$

For each input there is one and only one output so this is a function. What about the domain? Are there any numbers that we cannot use for x? Yes, there are. If x = 3 or x = -3, this expression would be undefined. Therefore, the domain is all x except 3 or -3. As for the range, we already know that y can never be zero. Therefore the range is all y except 0. Since there are points where the function is undefined, this is not continuous. This function has asymptotes, but you may have to adjust the window carefully on your calculator to see them. To help you locate the asymptotes, note where the domain and range are undefined. The numbers 3 and -3 are not part of the domain. In fact, x = 3 and x = -3 are vertical asymptotes: the graph of the function gets close to these vertical lines but never crosses them. Also, y = 0 is not in the range. The graph of the function gets close to this horizontal line without crossing it, so y = 0 is a horizontal asymptote. Here is the graph of this function.

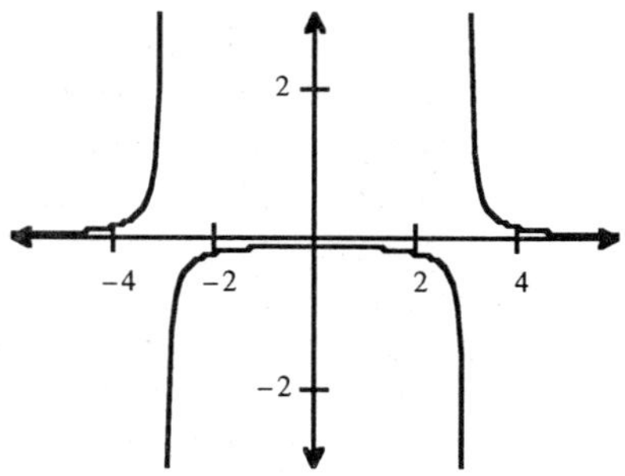

Now that you have finished three units in Algebra 2, stop and ask yourself some questions. Are you doing as well as you would like to be doing? Are you working as hard and diligently as you could be? What can you do to improve your learning? Is your tool kit up-to-date and readable? And the most important question of all: is the mathematics making sense? As hard as it may be for you to believe at times, the mathematics you are learning is not random. Methods progress in logical sequences, with reasons for everything that is done. If the work you do seems random, or if you seem to always "guess" as to what you should do next, then you are not truly understanding. If this is the case, you need to focus on the big picture and on making connections rather than small details. Already we have seen numerous connections, the most recent being that exponential functions and geometric sequences are really the same thing. As you go through the year, more and more items will connect to other items. If you find that you still need help in these matters and this guide isn't helping enough, be sure to contact your teacher. He/she should be more than willing to sit down with you and discuss the course so far.

UNIT 3
MORE TO TRY

Simplify the following as much as possible.

1. $(3x^{-2})^{-5}(4x^3)$ 2. $\dfrac{3x^{-4}y^3}{x^3y^{-5}}$ 3. $\dfrac{(9x^2y^4)^{-2}}{3x^{-7}y^8}$ 4. $\left[\dfrac{-27x^{-12}y^8z}{(-3x^4y^2z)^{-2}}\right]^0$

5. $\dfrac{-8x^{-7}a^5y^{-3}}{(2x^{3.5}a^{-2.5}y^{1.5})^{-2}}$ 6. $\left(16a^8b^{12}\right)^{3/4}$ 7. $\dfrac{a^{2/3}b^{-3/4}c^{7/8}}{a^{-1/3}b^{1/4}c^{1/8}}$ 8. $\dfrac{144^{1/2}x^{-3}}{(16^{3/4}x^7)^0}$

9. If $f(x) = 3\left(\frac{2}{3}\right)^x$, find f(4). 10. If $g(x) = -6x^7$, find g(-3). 11. If $h(x) = \frac{1}{4^x}$, find h(-2).

Solve the following for x. Be careful, show your work and check your answer.

12. $x^8 = 65536$ 13. $-5x^{-3} = \frac{25}{40}$ 14. $3^{5x} = 9^{x-1}$ 15. $16^{2x+1} = 8^{3-x}$

16. $3\cdot4^{x+3} = 3072$ 17. $\left[\dfrac{2^{x-1}}{3^{4x+3}}\right]^x = 1$ 18. $2(3x-5)^4 = 392$ 19. $\dfrac{2^{4x-1}}{2^{3x+2}} = 4$

Find the error in each of the following solutions. Then give the correct solution.

20.
$$\begin{aligned} 4(x+7)^6 &= 1392 \\ (x+7)^6 &= 348 \\ x+7 &= 58 \\ x &= 51 \end{aligned}$$

21.
$$\begin{aligned} 5^{4x+2} &= 10^{3x-1} \\ 5^{4x+2} &= 2\cdot5^{3x-1} \\ 4x+2 &= 2(3x-1) \\ 4x+2 &= 6x-1 \\ 3 &= 2x \\ x &= 1.5 \end{aligned}$$

Find all (if any) of the asymptotes for the functions and identify each as either vertical or horizontal.

22. $y = \frac{1}{x-3}$ 23. $f(x) = \frac{x}{x-3}$ 24. $g(x) = \frac{1}{x^2-1}$ 25. $y = 3\left(\frac{3}{4}\right)^x$

26. In seven seasons, Seta's son Stu sets sail for college. Seta hopes to have \$8000 saved up for his first year. She has \$5000 now and has found a bank that pays $7\frac{3}{4}$ % interest, compounded daily. At this rate, will she have the money she wants? If not, how much more does she need?

27. Eight years ago, I thought I was making a sound investment by buying \$1000 worth of "Pea Babies." Unfortunately, it hasn't turned out to be the investment I had hoped! My Pea Babies have depreciated steadily, each year losing 15% of their value. How much are they worth now? Justify your answer.

28. The new Bamo Super Ball has a rebound ratio of .97. If you dropped the ball from a height of 125 feet, how high will it bounce on the tenth bounce?

29. Fredrico's Bank will let you decide how often your interest will be computed, but with certain restrictions. If your interest is compounded yearly you can earn 8%. If your interest is compounded quarterly you earn $7\frac{7}{8}$ %. Monthly yields $7\frac{3}{4}$ % interest rate while weekly is computed with $7\frac{5}{8}$ %. If your interest is compounded daily you earn $7\frac{1}{2}$ %. What is the best deal? Justify your answer.

30. Investigate the equation $y = \left(\frac{3}{4}\right)^x + 4$

SAT PREP

1. If $-1 < t < 0$, which of the following statements must be true?

   a) $t^3 < t < t^2$ b) $t^2 < t^3 < t$ c) $t^2 < t < t^3$

   d) $t < t^3 < t^2$ e) $t < t^2 < t^3$

2. Without taking a single break, Merced hiked for 10 hours, up a mountain and back down by the same path. While hiking, she averaged 2 miles per hour on the way up and 3 miles per hour on the way down. How many miles was it from the base of the mountain to the top?

   a) 4 b) 6 c) 9 d) 12 e) 18

3. When a certain rectangle is folded in half, it forms two squares. If the perimeter of one of these squares is 28, what is the perimeter of the original rectangle?

   a) 30 b) 42 c) 49 d) 56 e) Cannot be determine from the information given.

4. A class of 50 girls and 60 boys sponsored a road rally race. If 60% of the girls and 50% of the boys participated in the road rally, what percent of the class participated in the road rally?

   a) 54.5% b) 55% c) 57.5% d) 88% e) 110%

5. The sum of four consecutive integers is s. In terms of s, which of the following is the smallest of these four integers?

   a) $\frac{s-6}{4}$ b) $\frac{s-4}{4}$ c) $\frac{s-3}{4}$ d) $\frac{s-2}{4}$ e) $\frac{s}{4}$

6. On a certain map, 30 miles is represented by one-half inch. On the same map, how many miles are represented by 2.5 inches?

7. How many of the first one hundred positive integers contain the digit 9?

8. The sum of n and n + 1 is greater than five but less than 15. If n is an integer, what is one possible value of n?

9. In the figure at right, ΔABC is a right triangle and $\frac{y}{6} = \frac{6}{x+y}$. What is the value of y?

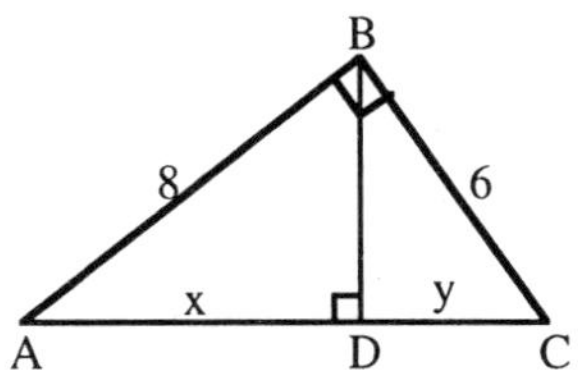

10. For three numbers a, b, and c, the average (arithmetic mean) is twice the median. If $a < b < c$, $a = 0$, and $c = kb$, what is the value of k?

UNIT 4

The Gateway Arch
PARABOLAS AND OTHER PARENT GRAPHS

UNIT 4
PARABOLAS AND OTHER PARENT GRAPHS

Things have been building--topics expanding--slowly broadening your awareness of mathematics. Next comes a very important unit where quite a bit of information will be introduced. It will be very important for you to look for (and see!) the connections to make the information manageable.

Let's start with a quick review. Suppose we have a function f(x) and we want to find the x-intercepts. The x-intercepts occur when y = 0 (you should **know** and **understand** this fact and it should be in your tool kit!) so we can find the x-intercepts by solving f(x) = 0. When f(x) is a quadratic function (as it often is in this unit) this means solving an equation which looks like $ax^2 + bx + c = 0$. As you probably remember there are several ways to do this. If the quadratic expression is factorable, we can factor and use the Zero Product Property to solve for x. Consider this first problem.

1. **What are the x-intercepts of $f(x) = 3x^2 - x - 2$?**

Solution.
As we just reviewed, the x-intercepts occur when $3x^2 - x - 2 = 0$. Can we factor to solve this? Sure! After a few guesses and checks (which is what factoring is, after all) we have

$$3x^2 - x - 2 = 0$$
$$(3x + 2)(x - 1) = 0$$

Using the Zero Product Property we write

$$3x + 2 = 0 \qquad\qquad x - 1 = 0$$
$$3x = -2 \qquad\qquad x = 1$$
$$x = -\frac{2}{3}$$

There are two x-intercepts. One at $\left(-\frac{2}{3}, 0\right)$, the other at (1, 0).

A note about factoring: what most people do, even people like me who has been doing math for more years then you've been alive, is try to factor a quadratic quickly in their head, but if they don't see the factors within a few **seconds**, they use something else. If you find you aren't a good "factorer" don't worry. So what do you use if the expression doesn't factor or you cannot factor it quickly? Use the Quadratic Formula. Even if you have the formula memorized and can use it, it is probably wise to have this formula <u>with</u> an example in your tool kit. Remember: this formula works with <u>every</u> quadratic equation, even if it does factor. Of course this means you must know what a quadratic equation is! Just to remind you, a quadratic equation is an equation in which the degree is 2 (the highest exponent is 2), and as we've seen, looks like $ax^2 + bx + c = 0$. The formula says

$$\text{If } ax^2 + bx + c = 0, \text{ then } x = \frac{-b \pm \sqrt{b^2 - 4ac}}{2a}$$

Some common errors students make when using this formula are: leaving off a negative sign, not putting the expression "all over" 2a, and forgetting that it is -(4ac). Let's look at an example.

2. **Find the x-intercepts of $f(x) = -4x^2 - 6x + 1$.**

Solution.
To find the x-intercepts of this function we must solve the equation $-4x^2 - 6x + 1 = 0$. Here (careful now) $a = -4$, $b = -6$, and $c = 1$. (b is -6 since the equation can be written as $-4x^2$ **+ -6x** $+ 1 = 0$.) Using the formula,

$$x = \frac{-(-6) \pm \sqrt{(-6)^2 - 4(-4)(1)}}{2(-4)}$$

Be very careful when simplifying! Work out and show all steps because it is very easy to make a careless mistake.

$$x = \frac{6 \pm \sqrt{36 - 4(-4)}}{-8} = \frac{6 \pm \sqrt{36 + 16}}{-8} = \frac{6 \pm \sqrt{52}}{-8}$$

Unless your teacher asks you to simplify further, **DON'T**. This one can be simplified further.

$$x = \frac{6 \pm \sqrt{4 \cdot 13}}{-8} = \frac{6 \pm 2\sqrt{13}}{-8} = \frac{2\left(3 \pm \sqrt{13}\right)}{-8} = \frac{3 \pm \sqrt{13}}{-4}$$

Your teacher may ask for a decimal approximation or you may have to plot the point. It is hard to plot the point $\left(\frac{3 \pm \sqrt{13}}{-4}, 0\right)$ without an estimate of the expression. To find the decimal approximation, you have two numbers to calculate: $\frac{3 + \sqrt{13}}{-4}$ and $\frac{3 - \sqrt{13}}{-4}$. Be careful how you enter this into your calculator. Remember that a fraction bar is a divide sign and also implies parentheses: $\left(3 + \sqrt{13}\right) \div -4$ and $\left(3 - \sqrt{13}\right) \div -4$. These become approximately (-1.65, 0) and (.15, 0). Verify this yourself! And be sure you have notes about x-intercepts and the quadratic formula in your tool kit.

Next is the heart of the unit: parent graphs. As you probably discovered, all parabolas have the same general shape. Sometimes they are narrower, as with $y = 4x^2$ shown at right, and sometimes they are wider, as shown with $y = 0.1x^2$. A parabola can even be upside down, but it is still a parabola. Changing the equation slightly changes the shape and moves it.

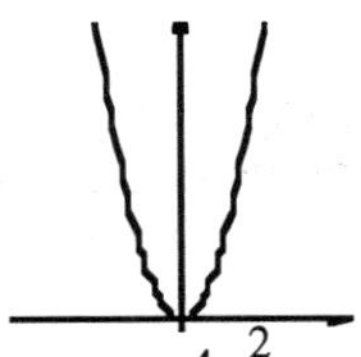
$y = 4x^2$

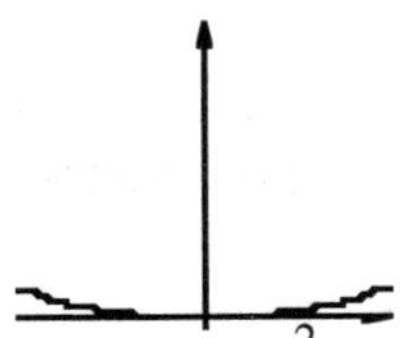
$y = .1x^2$

The graph at right shows the graph of $f(x) = x^2$, the parent of all parabolas. From this graph, you should be able to quickly sketch the graphs of $g(x) = (x + 2)^2$, $h(x) = (x - 4)^2$, $j(x) = x^2 + 1$, and $k(x) = x^2 - 6$. Functions g and h are shifted left and right respectively. Functions j and k are shifted up and down respectively. All four graphs are shown below. Once you understand this (graph them yourself to be sure you understand!) you should be able to combine them. In particular, you should be able to graph an equation in the form $y = a(x + h)^2 + k$, stating what each letter will do to the graph. Try different values for a, h, and k yourself to be sure that you can predict the effect that each of these numbers has on the graph. That's why $f(x) = x^2$ is the parent. You only need this one graph, and from it we can get all others just by moving, stretching and compressing.

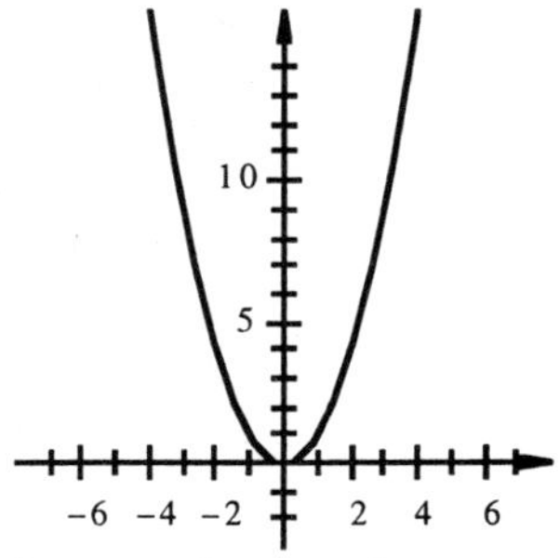

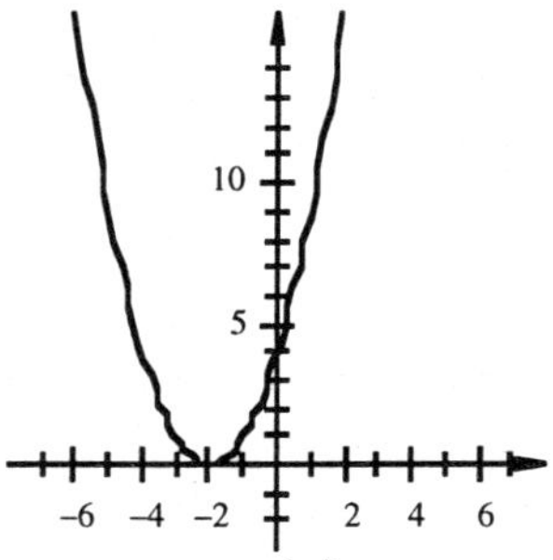

g(x)

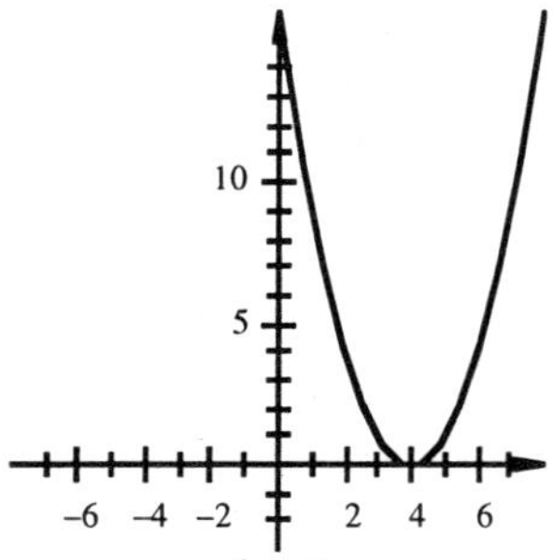

h(x)

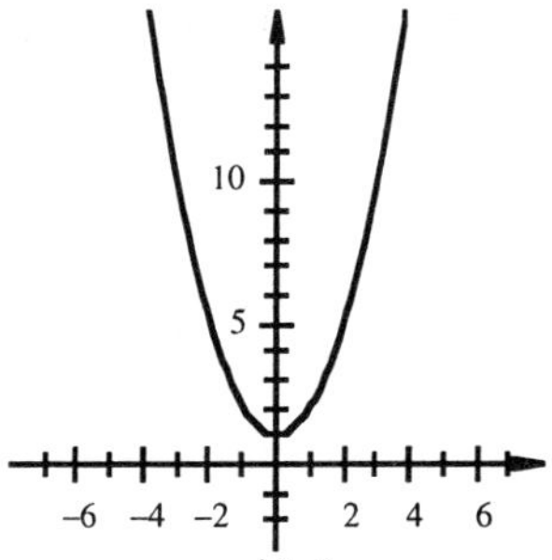

j(x)

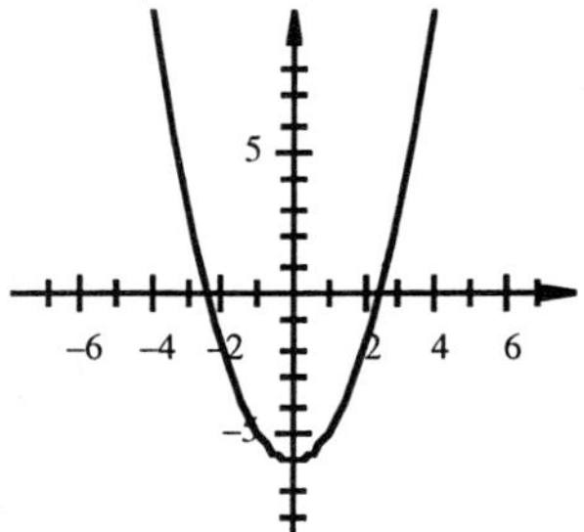

k(x)

What you must realize (and understand) is that this is true for all functions, not just parabolas. Consider this next example.

3. Suppose f(x) is shown at right.

a) What does f(x) + 3 look like? Graph it.

b) Graph f(x) - 2.

c) Graph f(x - 1).

d) Graph f(x + 3).

e) Graph 3f(x).

f) Graph $\frac{1}{2} \cdot f(x)$.

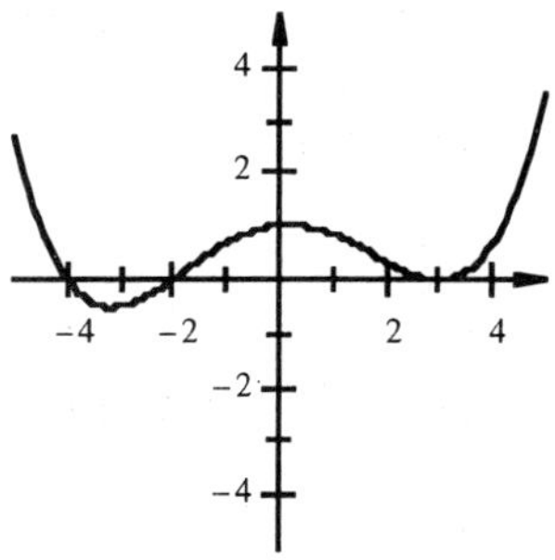

>>Solution appears on the next page.>>

Solution.

Each time we alter the equation slightly, the graph is changed. Even though we have no idea what the equation of this function is, we can still move it around. Think of the big picture when moving graphs. In part (a) we want to find out what f(x) + 3 looks like. Remember that f(x) represents the range or y values. Therefore this says "the y values, plus 3." Adding three to all the y values will lift the graph up three units. This is shown at right. Notice that the shape of the graph is exactly as the original, just shifted up three units. Check this by comparing the y-intercepts.

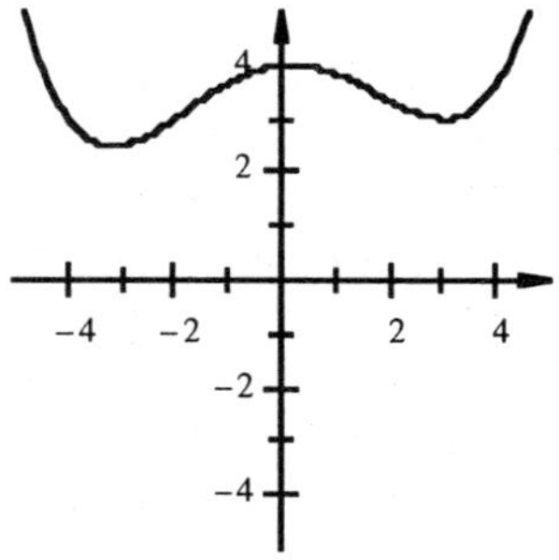

You can probably easily do part (b) now. If f(x) + 3 shifts the graph up three units, then f(x) - 2 will shift the graph down two units. Again compare the y-intercept in the graph below to the original. Using the y-intercept or the x-intercepts to help you graph is the smart way to create a graph. These points, as well as a vertex and other key points, are often called **locator points** because, as the name implies, they help you locate the position of the graph.

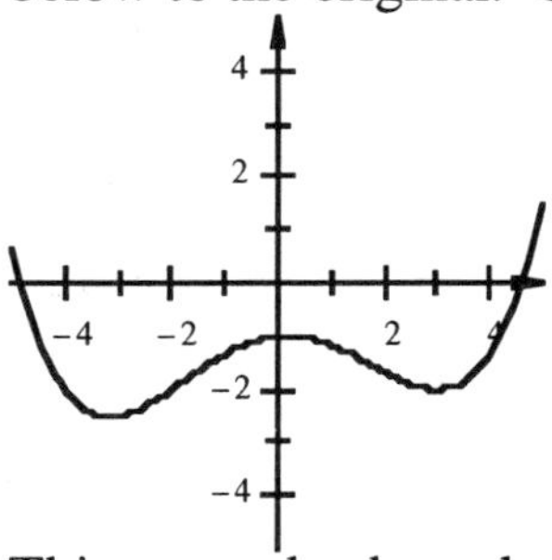

So what happens when the change is made within the parentheses as in parts (c) and (d)? Here we are making the shift with the x coordinates, thus the graph will move left or right. But, be careful! In part (c), the graph of f(x - 1) is shifted to the **right**. This seems backwards, but it is correct. What I do to help me remember this is I think of the x-intercepts. What is the x-intercept for y = x - 1? Solve x - 1 = 0. It is + 1, or think of it as one unit to the right. What is the x-intercept for y = x + 3? Solve x + 3 = 0. It is - 3, or three units to the left. If this doesn't help you remember which way to move, try to come up with your own method. Whether you use this idea or come up with one of your own, make sure you have the important information in your tool kit.

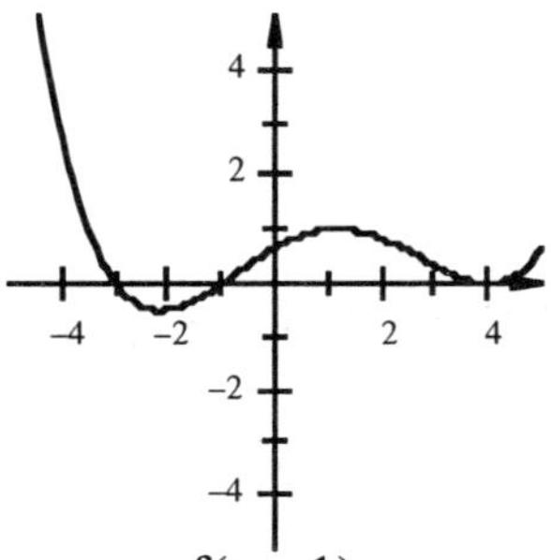

f(x - 1)

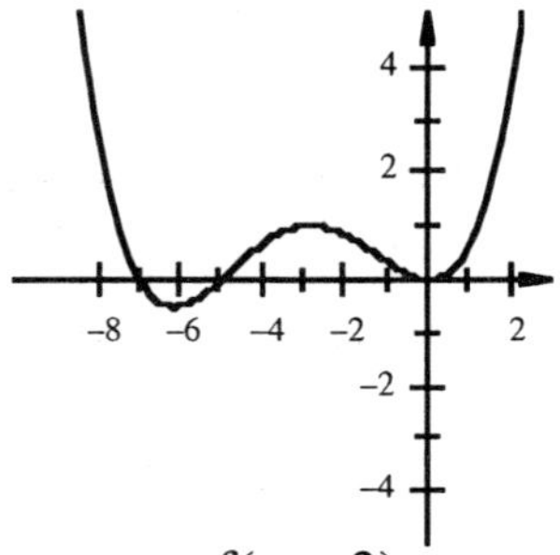

f(x + 3)

Multiplying f(x) by a number, as in parts (e) and (f), is what gives students the most trouble. Looking at some key points will help. To begin with, at the x-intercepts, f(x) = 0 so multiplying by 3 or $\frac{1}{2}$ won't change anything: $3 \cdot 0 = 0$ and $0 \cdot \frac{1}{2} = 0$. Therefore, the x-intercepts do not change at all. What about the y-intercept? In the original graph, the y-intercept is 1, so f(x) = 1 there. Multiplying by 3 will raise that point three times as high, to the point (0, 3). Multiplying by $\frac{1}{2}$ changes the y-intercept to $\left(0, \frac{1}{2}\right)$. The larger the constant that you multiply by, the more stretched out the graph becomes. A small number flattens the graph.

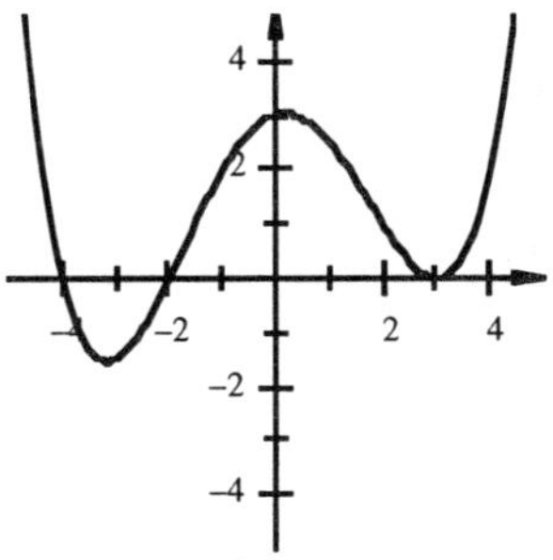

3f(x)

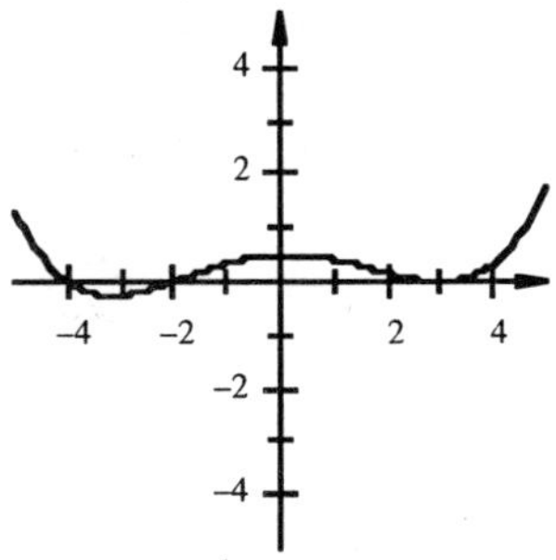

$\frac{1}{2}$f(x)

Look for the big picture and look for connections! I keep saying that and hopefully you can see why. Don't try to memorize separate rules for all the different parent graphs. Understand how and why the graphs change and that knowledge will work for <u>all</u> functions (and non-functions too when we get to that!) Try one more to see how you do. I will only print the solutions without explanations here. The explanations are similar to the ones given above.

4. The graph of $f(x) = |x|$ is shown at right.

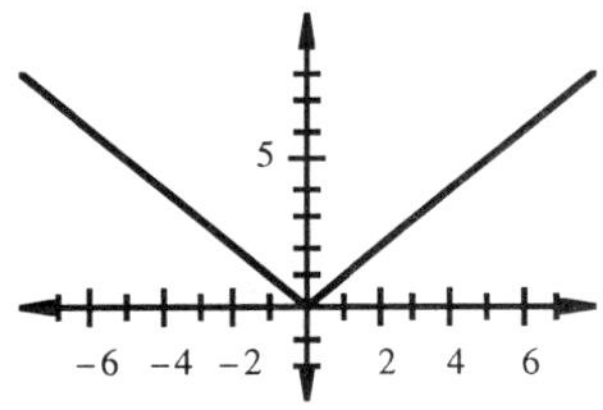

a) Graph $y = |x - 4|$.
b) Graph $y = |x| - 2$.
c) Graph $y = |x + 2| + 3$.
d) Graph $y = 3|x| + 1$

Solution.

a)

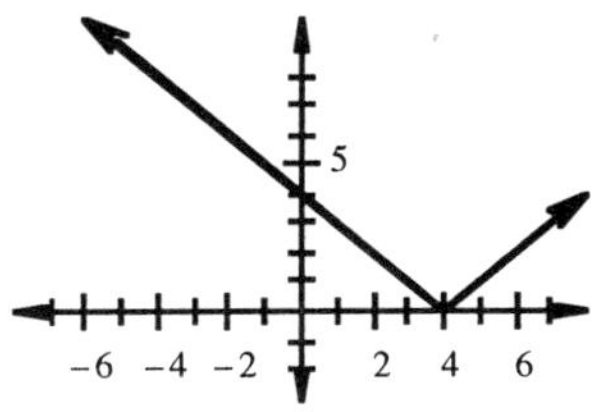

b)

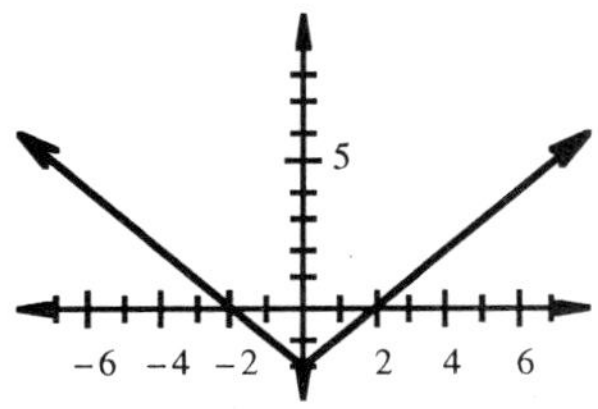

c)

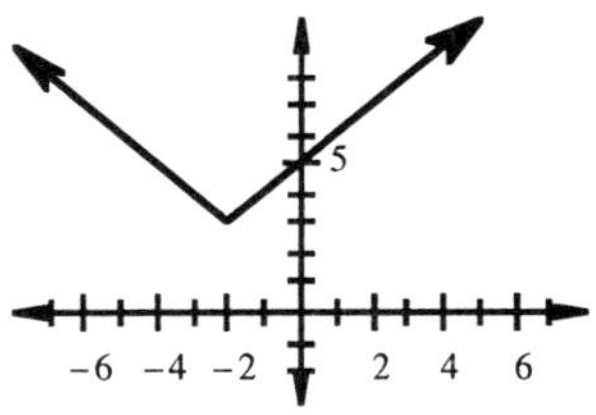

d)

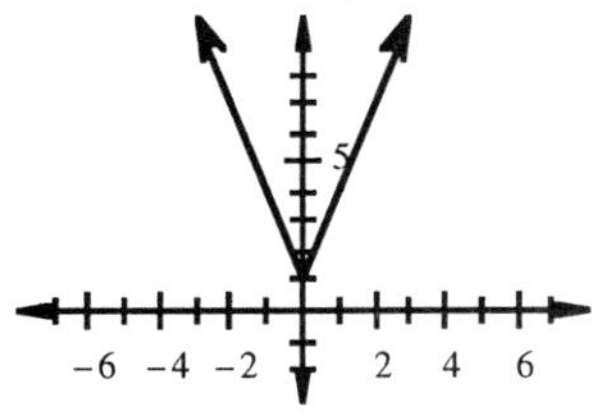

Once you have practiced moving graphs around by looking at the equation, then you are ready to write an equation to put a graph exactly where <u>you</u> want it. That's precisely what you are doing in the modeling problems such as PG-32, PG-33, PG-42 and the theme problem PG-151. For each of these you must write the equation of a parabola that will fit the data. Let's look at PG-42 first.

PG-42. While watering your outdoor plants, you notice that the water coming out of your garden hose follows a parabolic path. Write the equation of a parabola that describes the path of the water from the hose to the plant.

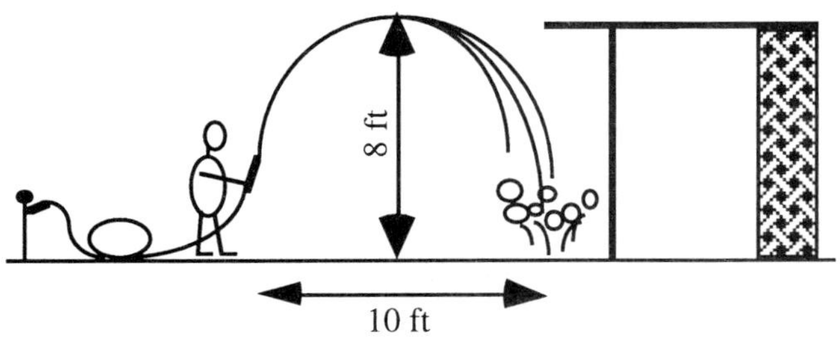

Solution.
We need to write the equation of a parabola that when graphed will look like the path of the water. One thing to realize with modeling questions is that the answer will vary depending on how you set up your axes. In this problem, for instance, we can set up the axes in several different positions.

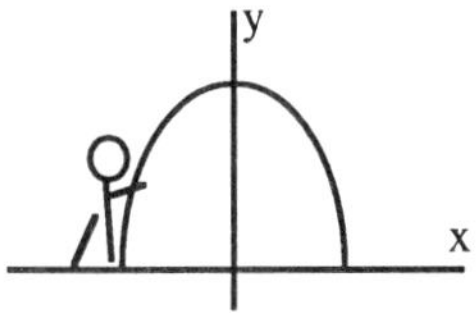

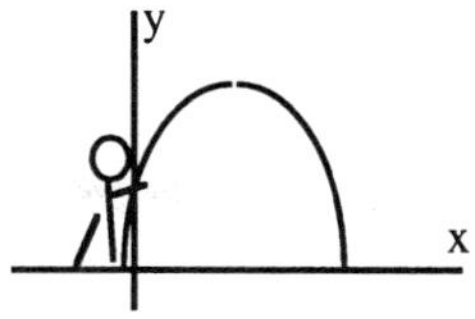

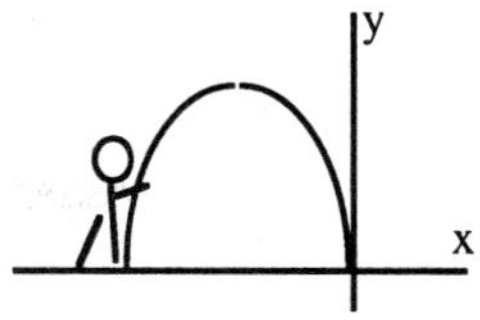

All of these are correct, but the answers will look different. You choose the approach that you think you will be able to do correctly. I will use the second one.

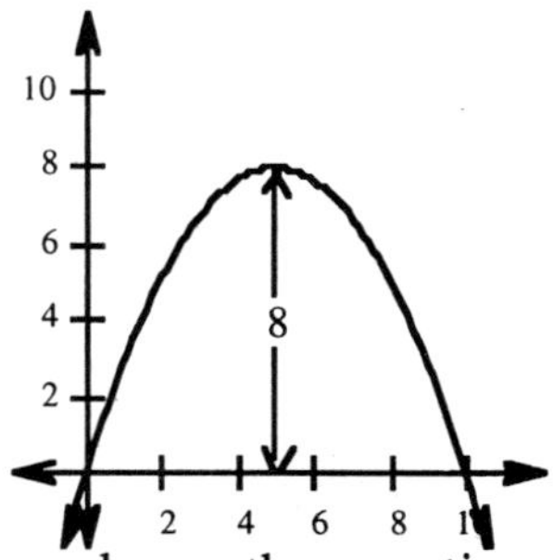

Remember, we are trying to write an equation $y = a(x + h)^2 + k$, whose graph looks like the one at right. We must find appropriate values for a, h and k. Do you know any of the variables already? We do know that a must be negative. Do you see why? (Hopefully so! This is important and should be in your tool kit. When $a < 0$, the parabola opens downward.) For this problem $a < 0$. What else can we figure out? This parabola is shifted to the right. By how much? The vertex (which in this case is the highest point) is shifted five units to the right. Which letter, a, h, or k, is responsible for the horizontal shift? h. So now we know the equation looks like $y = a(x - 5)^2 + k$, with a negative. What about k? The letter k tells us how the graph is shifted vertically. In this case the parabola has been shifted up eight units. Now our equation becomes $y = a(x - 5)^2 + 8$. All that's is left is to find a. Luckily, this parabola also goes through the point (0, 0) so let's use this point in the equation we have so far.

$$\begin{aligned} 0 &= a(0 - 5)^2 + 8 \\ 0 &= a(25) + 8 \\ 0 &= 25a + 8 \\ -8 &= 25a \\ a &= -\frac{8}{25} \end{aligned}$$

So we have done it! Our equation which models this data is

$$y = -\frac{8}{25}(x - 5)^2 + 8.$$

And do you know what is so great about these problems? You can check them! Just enter this equation into your graphing calculator and see if it looks like the graph we wanted. (Always check these! It is very easy to do with a graphing calculator and will alert you to any mistakes, careless or otherwise.)

Remember: if you used a different set up for this last problem, you would have found a different equation. (You did do the last problem before looking at the solution, right?) Check your equation by graphing it on your graphing calculator. If your graph has x-intercepts that are ten units apart and has a vertex eight units above the axis, you did it correctly.

Let's look at a new one.

5. In the neighborhood water balloon battle, Davis has developed a winning strategy. He has his home base situated five feet from an eight foot fence. Twenty-five feet away on the other side of the fence is his nemesis' camp. Davis, using a water balloon launcher, shoots his balloons so that they just miss the fence and land in his opponents' camp. Write an equation that when graphed, will model the trajectory (path) of the water balloon.

Solution.

As with many problems, the first thing to do is draw a sketch of the situation. The parabola shows the path the balloon will take, starting five feet from the fence and landing 25 feet past the fence.

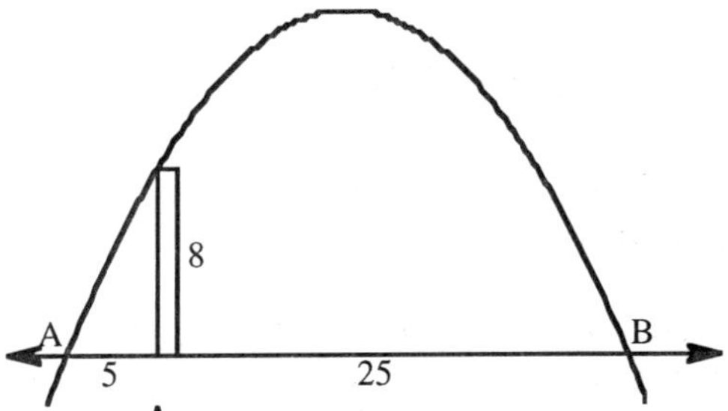

Now we must decide where to put the axes. Depending where you put them, your answer might be different than mine. I will set this one up with the fence on the y axis. The next thing I like to do is draw the parabola I'm trying to find the equation of and fill in any points I know that are on the parabola. I've done this in the second figure. This shows all the points we know just from the description of the problem. A <u>really</u> helpful point to find is the coordinates of the vertex. With such a path the high point of the parabola is the vertex (you should know that!) and it is half way between the x-intercepts. The total distance from A to B is 30 units, so half is 15. Fifteen units from point A is the point (10, 0). Now we know the equation looks like $y = a(x - 10)^2 + k$, with $a < 0$ again. Also, k must be greater than eight. (Do you know why $k > 8$?) We also know that the parabola goes through the points (0, 8), (-5, 0) and (25, 0). Let's use these points in the equation we have so far and see what else we can find.

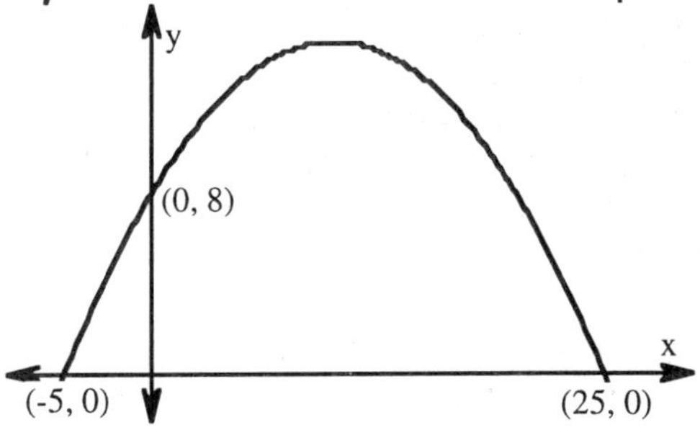

Using the point (0, 8) we can write

$$\begin{aligned} 8 &= a(0 - 10)^2 + k \\ 8 &= 100a + k \end{aligned}$$

Two variables are left, which tells us that we need another (different) equation with a and k to be able to solve this. Let's use the point (-5, 0).

$$\begin{aligned} 0 &= a(-5 - 10)^2 + k \\ 0 &= 225a + k \end{aligned}$$

That's our second equation. Now we have the following system to solve. Subtract the second equation.

$$\begin{array}{rcl} 8 &=& 100a + k \\ \underline{0} &\underline{=}& \underline{225a + k} \\ 8 &=& -125a \\ a &=& -\dfrac{8}{125} \end{array}$$

Substitute this value for the variable a back into one of the equations to find k.

$$8 = 100\left(-\frac{8}{125}\right) + k$$
$$8 = -\frac{32}{5} + k$$
$$k = \frac{72}{5}$$

The equation we have been looking for is $y = -\frac{8}{125}(x - 10)^2 + \frac{72}{5}$.

Remember to check your answer. Graph it with your graphing calculator and see if it looks like what you want.

These last two modeling problems illustrate something true of many math problems: they cannot be done in one step. Some students won't try a problem if they can't get the answer quickly and/or in their head. But most problems cannot be done that way. You must take smaller parts of the problem, solving them bit by bit. Start with the parts you know you can do. Solve them and then move on to another part. Gradually you will get more data until you reach the solution. All the work that I show in these examples is the work I must do to solve the problem, even with all of my years of math experience. Be sure that you show all of your work, too!

Before the test on this unit, redo the modeling problems in the book and in this guide to be sure you understand and have mastered modeling problems. Or better yet, if you've seen enough of them, write your own and solve it!

We have moved and stretched functions, investigated them again and again, so you know exactly what a function is, right? Can you recognize the graph of a function? This unit also asks you to look at two common graphs which **are not** functions, so you need to be clear about what a function really is! In words, a function is a relationship between x and y in which for each x there is one and only one y value paired with it. The graph of a function passes what is known as the **vertical line test**. This means a vertical line passes through the graph of a function once and only once, no matter where you draw the vertical line.

What sort of graph doesn't possess this property? The first seen in this unit is the graph of $x = y^2$. This equation looks like the equation of a parabola we know well: $y = x^2$, but with one big difference: the y variable is squared. To find the graph of $x = y^2$, you can't just use your graphing calculator. (Graphing calculators can only graph functions. With extra work on your part sometimes you can rewrite non-functions so that your calculator will accept them.) Let's graph this equation by making a table.

6. Fill in the table for $x = y^2$ and graph the results.

x	4	3	2	1	0	-1	-2
y	±2						

Solution.

For each value of x in the domain, there are two values for y, a positive one and a negative one.

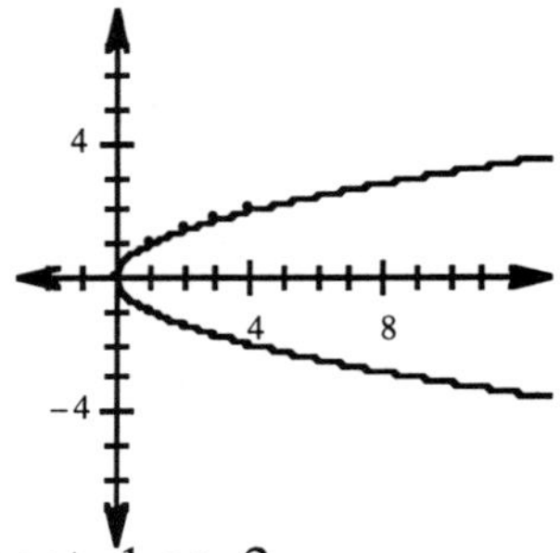

x	4	3	2	1	0	-1	-2
y	±2	±1.7	±1.4	±1	0	np	np

Note: zero is in the domain, but +0 and -0 are still just 0. Also, the y values for x = 3 and x = 2 are approximate. Why do we have "np" for the values of x = -1 and x = -2? Because squaring a number is always positive, so there is no number we could square to get -1 or -2

Examine this "sleeping parabola" (some books and teachers call these parabolas "sleeping" because they appear to be on their side) closely. Can you see that you can draw a vertical line that will cross the graph twice? This graph does not pass the vertical line test; so it is not a function. By the way, what are the domain and range for this relationship? (Domain: x values greater than or equal to 0, Range: y can be any real number.)

As you have probably guessed, even though $x = y^2$ is not a function, we can still move it around like the functions we have seen. What do you think the graph of $x = (y + 1)^2$ will look like? You can't just quickly enter it into your graphing calculator to check. You must make a table until you can see how changing the equation will change the graph. Let's do this in the next problem.

7. Graph each of the following on a separate set of axes. Then compare each graph to the graph of $x = y^2$.

a) $x = (y + 1)^2$
b) $x = (y - 3)^2$
c) $x = -y^2$
d) $x = y^2 + 2$
e) $x = y^2 - 3$

>>Solution appears on the next page.>>

8. In the equation $x = a(y - k)^2 + h$, explain completely what the letters a, k and h do.

Solutions.
I will make a table and graph part (a); the rest are left to you. I'll show the graph for parts (b) through (e) so that you can check but you should be able to create the tables yourself.

a) Remember to try several values for the input. Here is one possible set of points.

x	4	3	2	1	0	-1
y	1,-3	.7,-2.7	.4,-2.4	0, 2	-1	np

Note: the decimals are approximations.

Below are the graphs of the other parts.

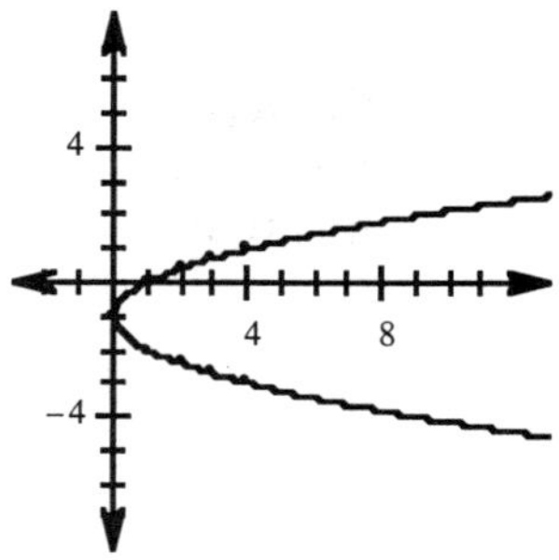

b)

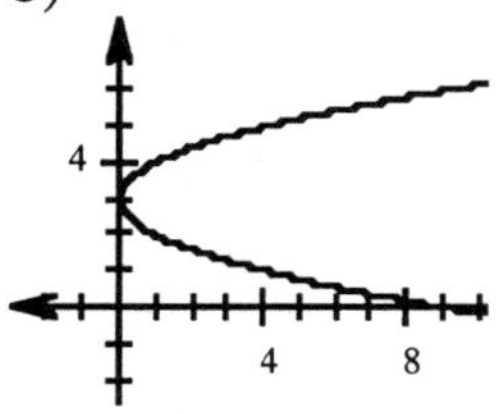

c)

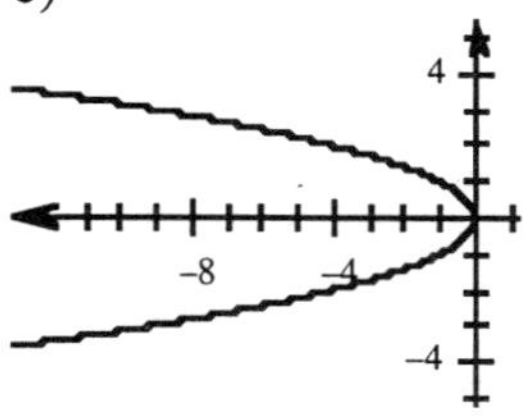

d)

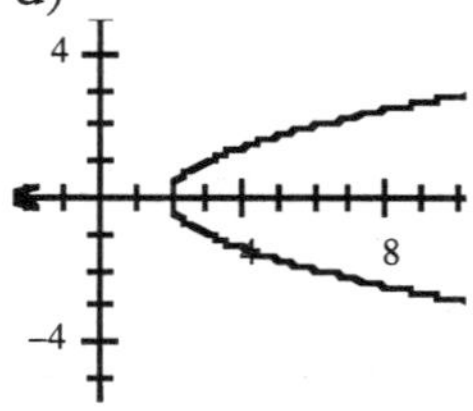

e)

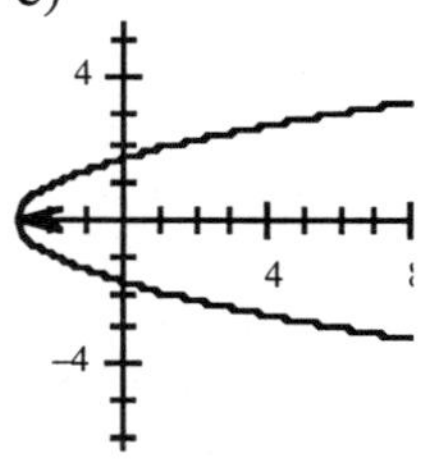

The letters move the sleeping parabola similarly to what we saw earlier. The variable a causes the parabola to open to the right when it is positive and to the left when it is negative. And as with the earlier parabolas, when a is small the parabola is wider. It is narrower when a is larger. The variable k slides the parabola up and down and h moves the parabola left or right.

Some students have trouble remember what h and k do. This might help. Some people will write $y = a(x - h)^2 + k$, as $y - k = a(x - h)^2$. Similarly, they write $x = a(y - k)^2 + h$ as $x - h = a(y - k)^2$. Then the letters move the parabola in the direction of the variable, x or y, that it is next to. That is, h moves the graph in the x direction (right or left) and k moves the graph in the y direction (up or down). Here, though you do have to remember which way--right or left, up or down--for each variable.

What is the other non-function we see in this unit? We graph $x^2 + y^2 = 9$ which is a circle with center (0, 0) and radius 3. Why isn't this a function? Make a quick sketch and check if it passes the vertical line test. The general equation of a circle is $(x - h)^2 + (y - k)^2 = r^2$, which is a circle with center (h, k) and radius r.

This unit also demonstrates a skill called **completing the square**. Be sure to read problems PG-46 and PG-47 carefully. Although you can memorize an algorithm for completing the square, remember the picture can help you do this method correctly. Below is an example (without the diagrams showing the "square" -- sorry, that takes lots of room.) typical of what you might see.

9. Give the vertex and sketch the graph of the functions below.

a) $f(x) = x^2 - 6x + 11$ **b)** $f(x) = -2x^2 - 16x - 37$

Solution.

If an equation of a parabola is written in vertex form then we can quickly sketch the graph. If you are used to finding the vertex of parabolas by averaging the x-intercepts, you might try doing that for these. There are no x-intercepts! This is a good time to be able to complete the square. Once completed, we will be able to easily read the vertex from the equation and then sketch the graph. As I mentioned, I'm not going to draw the accompanying diagram, but you may wish to do so. I think it is worthwhile to do while you are learning this procedure.

In part (a), we begin by writing the quadratic expression with a "hole" in it.

$$x^2 - 6x + 11 = x^2 - 6x + \underline{\quad} + 11$$

Our goal is to put the correct number in the space so that those first three terms form a perfect square. A perfect square fits this pattern:

$$(\Delta + O)^2 = \Delta^2 + 2\Delta O + O^2$$

where any number can replace the triangle and circle. In our example, triangle is x so if $2\Delta O = -6x$, then circle must be -3. That means $(-3)^2 = 9$ must go in the hole.

But wait! You can't just add 9 to the equation and expect it to stay balanced! If I'm going to add 9, I must also subtract 9 to keep the expression the same.

$$x^2 - 6x + 11 = x^2 - 6x + \mathbf{9 - 9} + 11$$

Now we group the first three terms and simplify.

$$\begin{aligned} x^2 - 6x + 11 &= (x^2 - 6x + 9) - 9 + 11 \\ &= (x - 3)^2 + 2 \end{aligned}$$

The vertex of this parabola is (3, 2). We also see that it opens upward (a is positive) and is the standard width (a = 1).

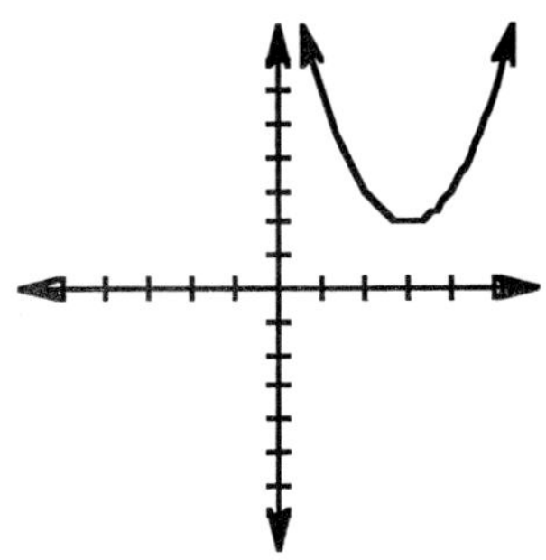

In part (b) we do the same thing <u>except</u> we have to deal with that -2 first. We "deal" with it by factoring it out of the first two terms.

$$-2x^2 - 16x - 37 = -2(x^2 + 8x + \underline{\quad}) - 37$$

Again I want to fill in the space with a number that will give me a perfect square trinomial. Here, 2ΔO = 8x and Δ = x, so O = 4. That means $O^2 = 16$. Again, I can't just put 16 in the equation--what must I do to keep this in balance? NOT 16! You see, once I put 16 in this space, it really is no longer a 16.

$$-2x^2 - 16x - 37 = -2(x^2 + 8x + \mathbf{16}) - \mathbf{?} - 37$$

Remember the distributive rule! That really is a -2(16) = -32, so I must subtract off -32 (or in other words <u>add 32)</u>.

$$-2x^2 - 16x - 37 = -2(x^2 + 8x + \mathbf{16}) - \mathbf{(-32)} - 37$$

Now we just simplify.

$$-2x^2 - 16x - 37 = -2(x + 4)^2 - 5$$

The vertex is (-4, -5) and this parabola opens downward. It is also slightly narrower than the first parabola (a = 2).

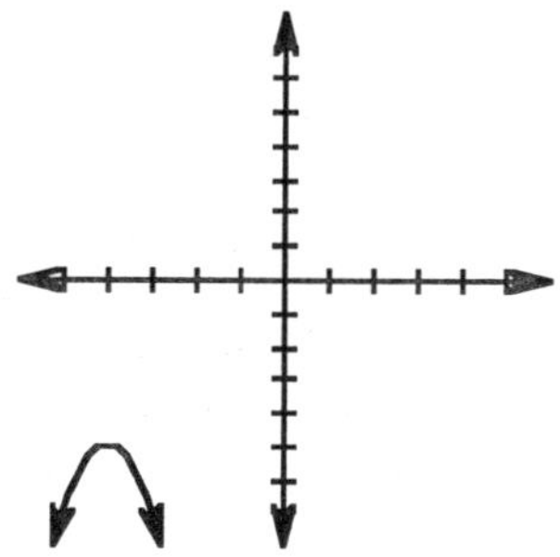

Note: if the Δ and O pattern I showed above doesn't help you understand this method, don't use it. It is just a pattern some students find useful when they are trying to make perfect square trinomials. Use the method that works best for you. In order to study and understand other curves in Unit 13, you will need to be able to complete the square efficiently.

This completes the review of Unit 4. Like I mentioned earlier: it is very important to look for the connections in this unit because there is quite a bit to understand.

UNIT 4
MORE TO TRY

Find the x- and y-intercepts of each of the following.

1. $y = x^2 + 4x + 3$
2. $y = x^2 + 5x - 6$
3. $y = 2x^2 - 7x - 4$
4. $y = -3x^2 - 10x + 8$
5. $y = 4x^2 + x - 5$
6. $y = 9x^2 - 12x + 4$
7. $y = 16x^2 - 25$
8. $y = 6x^2 + 2x - 7$
9. $y = -3x^2 + 10x + 5$
10. $y = 6x - 12$

Find the error in each of the following. Then find the correct solution to the problem.

11. Solve for x.

$$3x^2 + 6x + 1 = 0$$

$a = 3, b = 6, c = 1$

$$x = \frac{6 \pm \sqrt{6^2 - 4(3)(1)}}{2(3)}$$
$$= \frac{6 \pm \sqrt{36 - 12}}{6}$$
$$= \frac{6 \pm \sqrt{24}}{6}$$
$$= \frac{6 \pm 2\sqrt{6}}{6}$$
$$= \frac{3 \pm \sqrt{6}}{3}$$

12. Solve for x.

$$-2x^2 + 7x + 5 = 0$$

$a = -2, b = 7, c = 5$

$$x = \frac{-7 \pm \sqrt{7^2 - 4(-2)(5)}}{2(-2)}$$
$$= \frac{-7 \pm \sqrt{49 - 40}}{-4}$$
$$= \frac{-7 \pm \sqrt{9}}{-4}$$
$$= \frac{-7 \pm 3}{-4}$$
$$x = \frac{-4}{-4} = 1 \text{ or } x = \frac{-10}{-4} = 2.5$$

13. Solve for x.

$$8x^2 - 4x - 3 = 0$$

$a = 8, b = -4, c = -3$

$$x = 4 \pm \frac{\sqrt{16 + 96}}{16}$$
$$= 4 \pm \frac{\sqrt{112}}{16}$$
$$= 4 \pm \frac{4\sqrt{7}}{16}$$
$$= 4 \pm \frac{\sqrt{7}}{4}$$

For each of the following, state whether or not it is a function. If it isn't a function, explain why not.

14.

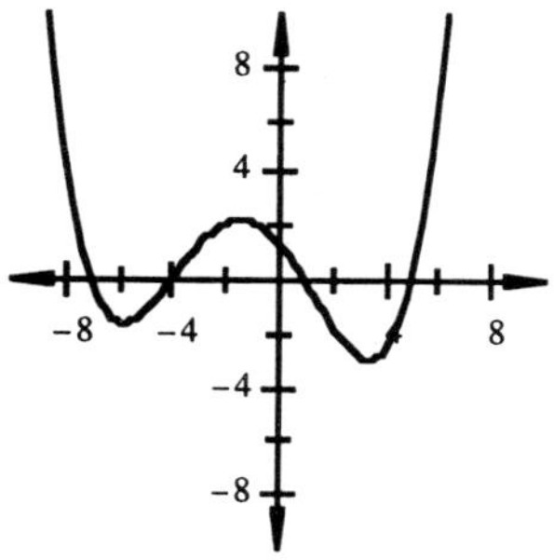

15.

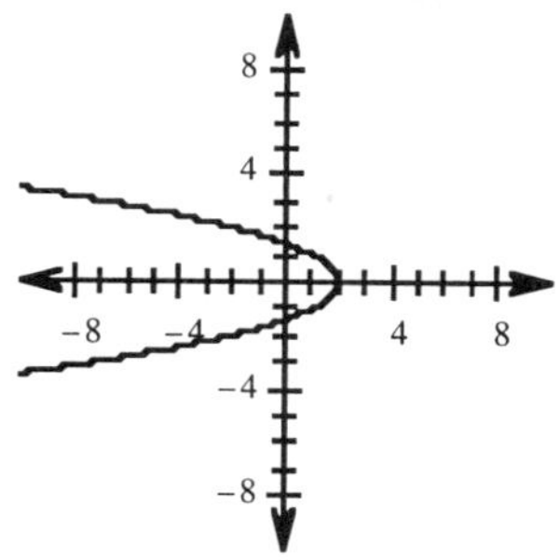

16. $y = 7 \pm \sqrt{9 - x^2}$

17. $y = 3(x - 4)^2$

Sketch the graphs of each of the following. Be sure to include any key and/or locator points.

18. $y = (x - 5)^2$

19. $y = -x^2 - 6$

20. $y = -\frac{1}{3}(x + 4)^2 + 7$

21. $y = |x + 5| - 2$

22. $y = \frac{1}{x + 1} + 10$

23. $y = 2^x - 8$

24. $y = -(x - 2)^3 + 1$

25. $y = \sqrt{x + 7}$

26. $y = \frac{1}{3^x} - 9$

27. $y = \frac{-2}{x + 3}$

For each situation, come up with the appropriate equation which will model it effectively.

28. When Twinkle Toes Tony kicked a football, it landed 100 feet from where he kicked it. It also reached a height of 125 feet. Write an equation that when graph will model the path of the ball from the moment it is kicked until it first bounces on the ground.

29. When some software companies develop software, they do it with "planned obsolescence" in mind. This means they plan on the sale of the software to rise, hit a point of maximum sales, then sales will drop and eventually stop. Thus the curve showing the number of sales over time is parabolic. Suppose a company plans on the "life span" of its product to be six months, with maximum sales reaching 1.5 million units. Write an equation which best fits this data.

30. A new skateboarder's ramp just arrived at Bungey's Family Fun Center. A cross-sectional view shows that the shape is parabolic. The sides are 12 feet high and 15 feet apart. Write an equation which, when graphed, will show the cross section of this ramp.

SAT PREP

1. If m is an integer, which of the following could **not** equal x^3?

   a) -27 b) 0 c) 1 d) 16 e) 64

2. If n is divided by 7 the remainder is 3. What is the remainder if 3n is divided by 7?

   a) 2 b) 3 c) 4 d) 5 e) 6

3. What is the slope of the line passing through the point (-3, -1) and the origin?

   a) -3 b) $-\frac{1}{3}$ c) 0 d) $\frac{1}{3}$ e) 3

4. If $x = 2y + 3$ and $3x = 7 - 4y$, what does x equal?

   a) -5 b) $-\frac{1}{5}$ c) $\frac{13}{5}$ d) $\frac{2}{3}$ e) 15

5. A bag contains a number of marbles of which 35 are blue, 16 are red and the rest are yellow. If the probability of selecting a yellow marble from the bag at random is $\frac{1}{4}$, how many yellow marbles are in the bag?

   a) 4 b) 17 c) 19 d) 41 e) 204

6. If $n > 0$ and $16x^2 + kx + 25 = (4x + n)^2$ for all values of x, what does $k - n$ equal?

   a) 0 b) 5 c) 35 d) 40 e) 80

7. A rectangular solid has two faces congruent to figure I at right and four faces congruent to figure II at right. What is the volume of the solid?

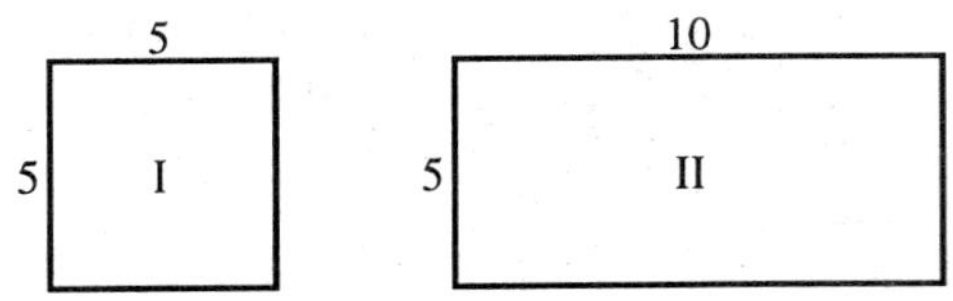

8. In the figure at right, PQ = QR. What is the x-coordinate of point Q?

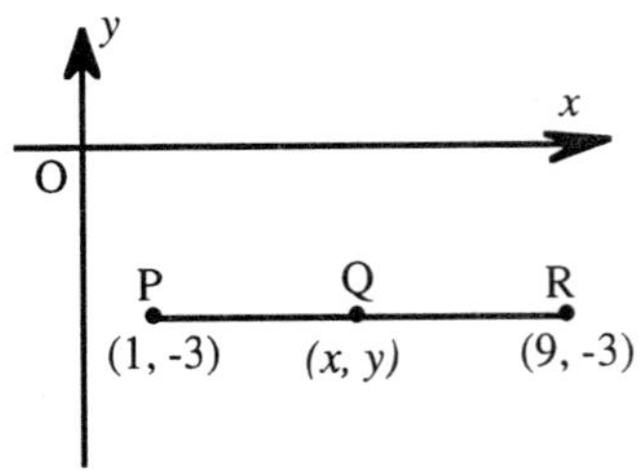

9. The time t, in hours, needed to produce u units of a product is given by the formula $t = ku + c$, where k and c are constants. If it takes 430 hours to produce 100 units and 840 hours to produce 200 units, what is the value of c?

10. In the figure at right, a square is inscribed in a circle. If the sides of the square measure $\sqrt{3}$ and the area of the circle is $c\pi$, what is the exact value of c?

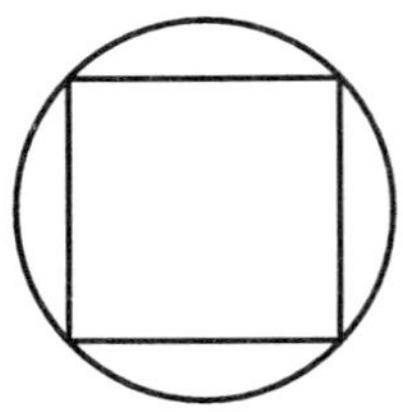

UNIT 5

The Toy Factory

LINEAR SYSTEMS AND MATRICES

UNIT 5
LINEAR SYSTEMS AND MATRICES

A big idea in this course is understanding what a solution is. In Algebra 1 a common goal is to have students understand that if $x = 5$ is a solution to the equation $3x + 2 = 17$, then substituting $x = 5$ into the equation will produce a true statement. Also, $x = 5$ is the only x value that will work in this equation.

In this course we take this idea further and we do so in this unit. We want you to know what a solution to an inequality is, what a solution to a system of equations or inequalities is, and what these solutions represent. In addition, we want to use what we know to solve some practical applications. To help us, we learn about matrices and matrix operations.

We start with inequalities. Many students might remember an elementary teacher telling them something like "Remember! The alligator wants to eat the bigger amount!" As corny as this seems, it is true and it works! Even for the mathematics we will be doing in this unit you can still count on the alligator! So, what goes in the blank here?

-5 ____ -10

Hopefully you answered >, greater than. Unfortunately, most of our problems involving inequalities will not be so simple. This course is algebra so we will not be working with numbers alone. We need variables! We begin by solving simple inequalities such as $x + 3 < 2$, which we do by solving as if the inequality were an equal sign.

$$\begin{aligned} x + 3 &< 2 \\ x &< -1 \end{aligned}$$

Also, be sure to **check** as we always do with equations. Checking inequalities, though, requires a little more thought. In this example we don't want to substitute $x = -1$ into the inequality. We must use something a little smaller (less than -1). (Another helpful hint: "less than" means "to the left of" on a number line and similarly, "greater than" means "to the right of" on a number line. So $-4 < -1$ because -4 is to the left of -1 on a number line.) Let's try -2 since it is less than (to the left of) -1.

$$\begin{aligned} x + 3 &< 2 \\ -2 + 3 &\overset{?}{<} 2 \\ 1 &< 2 \ \checkmark \end{aligned}$$

If we did try $x = -1$, we would find a problem.

$$\begin{aligned} x + 3 &< 2 \\ -1 + 3 &\overset{?}{<} 2 \\ 2 &< 2 \end{aligned}$$

which is **not** true! Two is not less than 2. If the problem had been $x + 3 \le 2$, then the statement: $2 \le 2$, is true. Remember that $\le$ (and $\ge$) means less than (greater than) **or** equal to. So $2 \le 2$ (and $2 \ge 2$) are both true statements since 2 is equal to 2.

Let's try a slightly harder one. Remember: try it on your own first.

1. Solve the following inequality and graph the solution: -3x + 7 > 5.

Solution.
As before, solve for x as if the inequality sign were just an equal sign.

$$\begin{aligned} -3x + 7 &> 5 \\ -7 \quad &\;\; -7 \\ -3x &> -2 \end{aligned}$$

Next we will divide by -3, but there is something else that must be done when we multiply or divide by a negative number to keep the statement true. Do you remember? We must reverse the inequality symbol.

$$\frac{-3x}{-3} < \frac{-2}{-3}$$

$$x < \frac{2}{3}$$

The graph of this solution is a number line graph. We want all x's less than (to the left of) $\frac{2}{3}$.

The fraction $\frac{2}{3}$, however, is not part of the solution so we keep an open circle at this point.

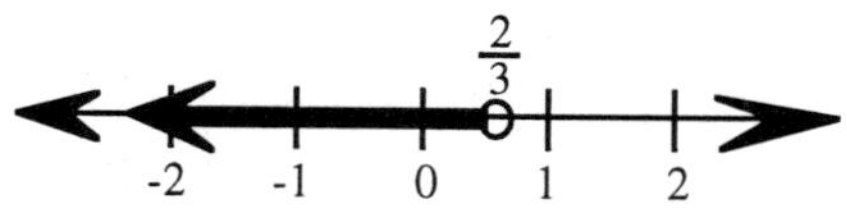

Be sure you have included the rule: "When multiplying or dividing both sides of an inequality by a negative number, reverse the direction or the inequality symbol" in your tool kit. You should also be able to demonstrate this rule. As an example, consider the following true statement:

$$5 < 10$$

Add or subtract a number from both sides and it is still true.

$$5 + 3 < 10 + 3$$
$$8 < 13$$

Or,

$$5 - 6 < 10 - 6$$
$$-1 < 4.$$

If we multiply or divide by a positive number, the statement is still true.

$$3 \cdot 5 < 3 \cdot 10$$
$$15 < 30$$

Or

$$5 \div 5 < 10 \div 5$$
$$1 < 2$$

But! Try this same thing with a negative number and there is trouble!

$$5 < 10$$
$$-3 \cdot 5 \qquad -3 \cdot 10$$
$$-15 \nless -30$$

Or

$$5 < 10$$
$$5 \div -5 \qquad 10 \div -5$$
$$-1 \nless -2$$

To make each of these false statements true again, we must reverse the inequality: -15 > -30 and -1 > -2. Don't let this small step lead you to making careless errors! Put a note in your tool kit.

So what would you do with something like y > 2x + 3? If you're not sure, go back to a simpler, similar case: what would you do with y = 2x + 3? Graph it! What do we get when we graph this equation? A line. And all the points on this line make the equation true. Try each of the these points in the equation to show that they make the equation true, and show that they are on the line: (0, 3), (1, 5), (-2, -1).

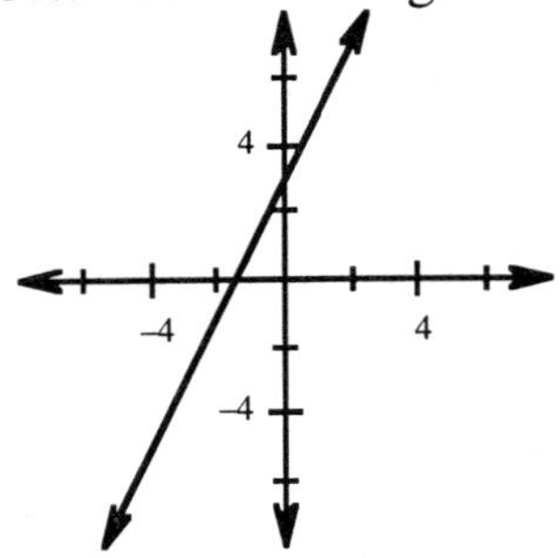

$$3 \stackrel{?}{=} 2(0) + 3 \qquad 5 \stackrel{?}{=} 2(1) + 3 \qquad -1 \stackrel{?}{=} 2(-2) + 3$$
$$3 = 3 \qquad 5 = 5 \qquad -1 = -1$$

They all are true. How many points could make this equation true? Infinitely many. Every point on the line makes the equation true, even the messy ones like $\left(\frac{3}{7}, \frac{27}{7}\right)$. How many points would make this equation false? Every point <u>OFF</u> the line -- that's infinitely many points as well. For instance, the point (5, 2) is not on the line and

$$2 \stackrel{?}{=} 2(5) + 3$$
$$1 \neq 13$$

This is one of the things that is so nice about a graph: it quickly shows information about infinitely many points! But as you can also see, you don't need the graph to know whether or not a point is on the graph.

Now let's take the next step. Certainly y > 2x + 3 is similar to y = 2x + 3. In fact to graph this inequality, we start with the graph of y = 2x + 3. We want all the points whose y-coordinate is larger than 2 times the x-coordinate, plus 3. For instance, check the point (0, 0). Is 0 > 2(0) + 3? No, 0 is not greater than 3, so that's not one of the points in the solution. What about (-4, 5)?

$$5 \stackrel{?}{>} 2(-4) + 3$$
$$5 > -5$$

Yes! So what does this tell us? We checked a point below the line and it does not make the inequality true. A point above the line, (-4, 5), did make the inequality true. Try another point above the line, any point. (DO IT!) Did it make the inequality true or false? It should! **<u>Every</u>** point above the line will make the inequality true. This means there are infinitely many of them again so we can't list all the points. To show this, we shade the graph above the line.

What about a point on the line? The line is the <u>boundary</u>. Try the point (0, 3) in the inequality.

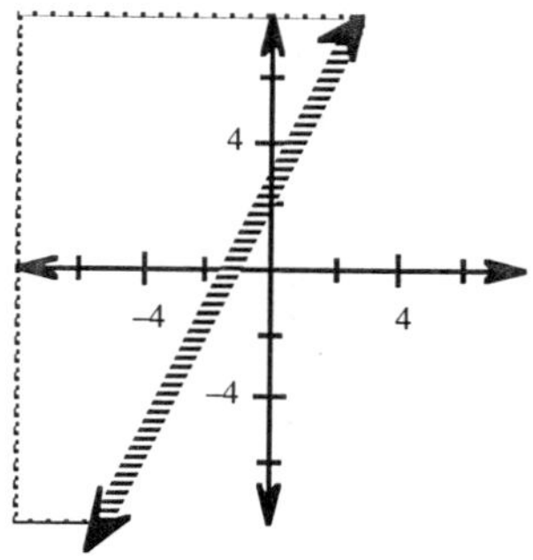

$$3 \overset{?}{>} 2(0) + 3$$
$$3 > 3$$

Not true! Try two more points on the line. (DO IT!) Both of the points you choose should make the inequality false. This tells us that points on the line are not included in the solution. We show this by making the line dotted, thus not including it in the solution. As you have probably guessed, if the problem had been to graph $y \geq 2x + 3$, we would include the line and would make the line solid in the graph.

Once you understand this process of graphing the solutions of inequalities (graph as if it were an equal sign, then determine which side to be shaded and whether the boundary is solid or dashed), you can graph the solution of almost any inequality. Consider this next example.

2. Graph the solution to the inequality $y \leq 3(x - 2)^2 - 1$.

Solution.

We can graph $y = 3(x - 2)^2 - 1$ quickly by realizing it is a parabola with vertex at (2, -1) (locator point), it opens up, and is narrow.

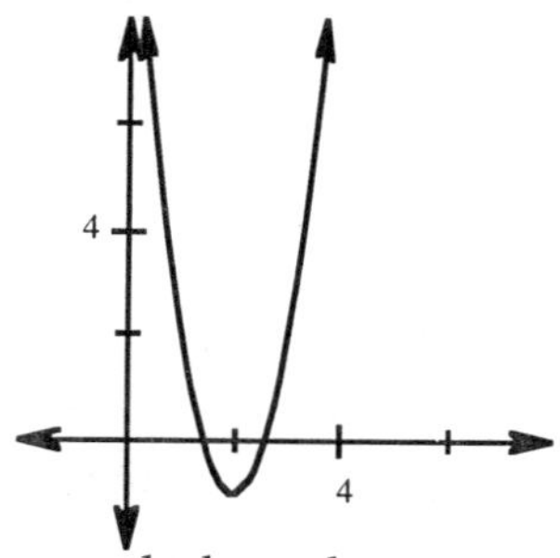

The parabola is the boundary between the shaded and the unshaded part of the graph. In this case, you can think of the two parts being the inside and the outside of the parabola. Choose two points from each region and substitute them into the original inequality. Which points make the inequality true? Which make it false? You should have discovered that points outside the parabola make the inequality true. For instance, the point (2, -4) is outside the parabola and

$$-4 \overset{?}{\leq} 3(2 - 2)^2 - 1$$
$$-4 \overset{?}{\leq} 3(0)^2 - 1$$
$$-4 \overset{?}{\leq} 0 - 1$$
$$-4 \leq -1, \text{ true!}$$

We shade the points that make the inequality true so here we shade outside the graph. We <u>do</u> include the parabola -- why? Because we want $y \leq$ (less than or equal to) $3(x - 2)^2 - 1$.

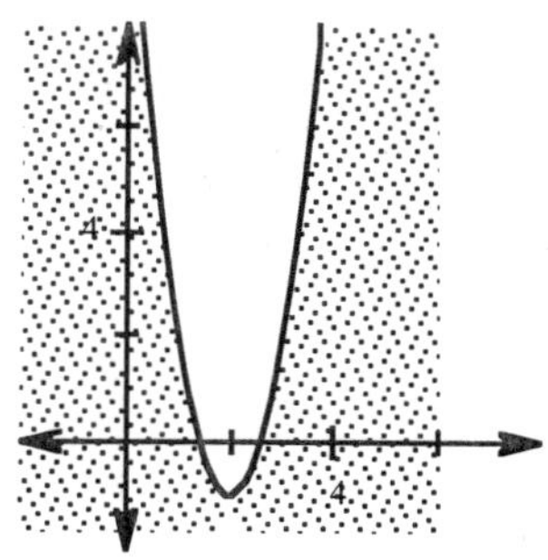

Note: Sometimes with quick sketch of the parabola, you may not be able to recognize whether a point is inside or outside. For instance, in the previous example, is the point (1, 1) inside or outside? If your are not sure about a point, DON'T use that point! I used the point (2, -4) which is directly under the vertex because I can be sure it is outside of the parabola.

Now that we can graph inequalities, we can consider combinations of inequalities to create systems of inequalities. Try this next example.

3. Graph the system: $\begin{cases} y < \frac{-2}{3}x + 5 \\ y > x^2 - 1 \end{cases}$

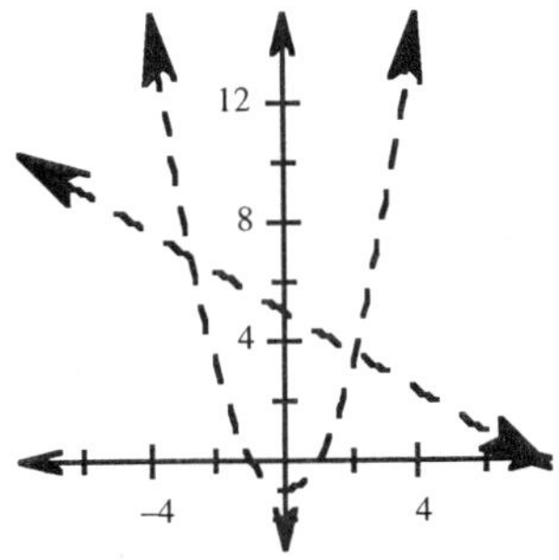

Solution.
As with all the other inequalities, first treat each of these as if it had an equal sign, and graph the two on the same set of axes. Each will be a dotted curve/line -- why? Because each inequality is not "or equal to". Next, using different colored pens, or by shading in different directions, we will determine the shading for each one. Do these one at a time. First, let's start with the line. Using the point (0, 0) as a test point (always try to choose a point that will make things easy for you), substitute it into the first inequality to see if it makes a true or false statement.

$$0 \overset{?}{<} \frac{-2}{3}(0) + 5$$
$$0 < 5, \text{ true!}$$

This means we will shade below the line. Note: don't make this shading very dark. It may not all be part of the final solution.

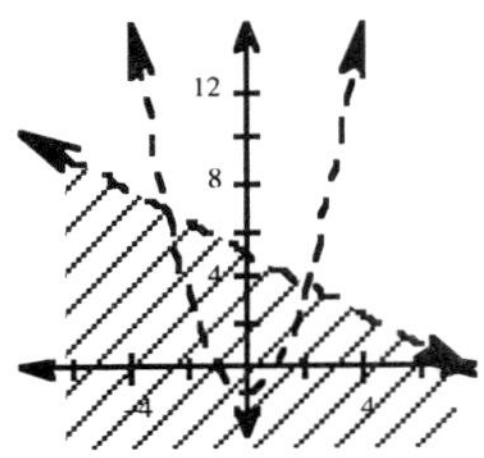

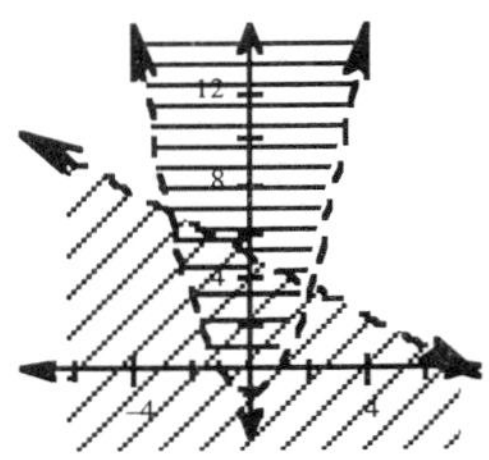

Next we will determine whether we shade the inside or the outside of the parabola. The point (0, 0) will work as a good test point for this inequality as well. (Remember you can use any point inside or outside the parabola. Just don't choose a point **on** the parabola!)

$$0 \overset{?}{>} (0)^2 - 1$$
$$0 > -1, \text{ true!}$$

We shade inside the parabola as shown above, right.

We want to find the region that makes both inequalities true at the same time. This happens where the two shaded regions overlap. That is the solution to the system. Here, the regions overlap below the line and inside the parabola.

It may not be required, but it is always nice to redraw the picture making the solution to the system very clear. I've done this at right.

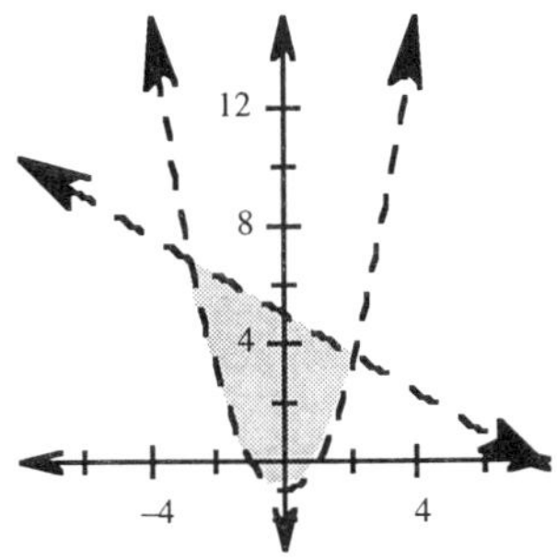

This next example is a good indicator of your understanding. Try it and then read the solution and comments below to evaluate your level of understanding. And don't underestimate yourself by thinking "This is too easy; I must be doing something wrong." Trust yourself. Remember also, if you try and get it wrong, you can learn something. If you don't try at all, you'll never know.

4. **The graph of**

$$y = .01x^4 - .01x^3 - .21x^2 - .01x + .2$$

is shown at right. Graph

$$y < .01x^4 - .01x^3 - .21x^2 - .01x + .2$$

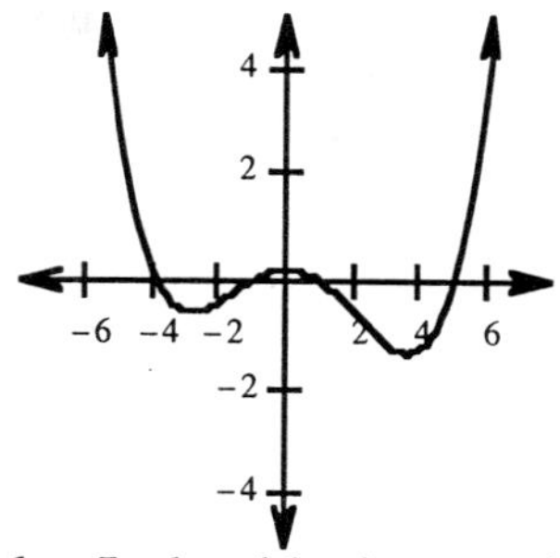

Solution.
If you understand graphing inequalities, this problem is simple. It should take very little work. The function expressions are the same. The only change is the equal sign to an inequality symbol. This means the same curve is the boundary, but it is dashed (because it is <, not ≤). All we need to do is determine which side to shade. We can try using any point, not on the curve, to use as a test point. To keep it simple, I'll use (0, 0).

$$0 \overset{?}{<} .01(0)^4 - .01(0)^3 - .21(0)^2 - .01(0) + .2$$
$$0 \overset{?}{<} 0 - 0 - 0 - 0 + .2$$
$$0 < .2, \text{ true!}$$

So we shade below the curve. All we need to do is trace the curve, dashed, and add shading below the curve. The solution is shown below.

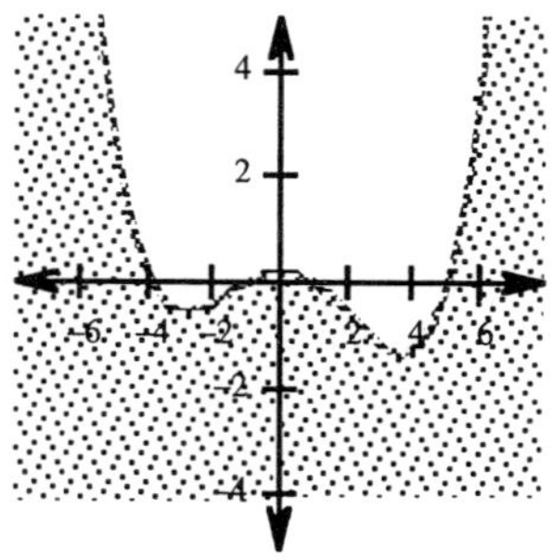

Here is another question that is a good check for understanding.

5. **At right is the graph of the inequality in the last example. The point A has coordinates $\left(-2\frac{17}{123}, 4\frac{27}{195}\right)$. Would substituting this point into the inequality**

$$y < .01x^4 - .01x^3 - .21x^2 - .01x + .2$$

make a true or false statement? Explain.

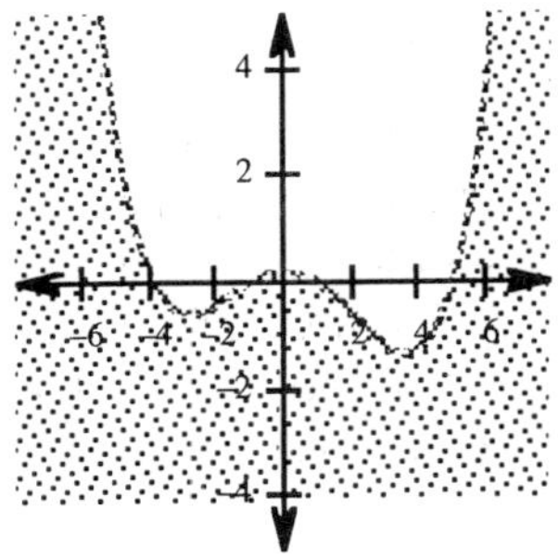

Solution.
If you think that the word "Explain" is the reason why I consider this problem to be a good "check for understanding" you are only partly right. Certainly, explanations add insight to your level of understanding. But the true reason this problem is a good check is because if you **truly** understand solutions of inequalities, this is a very easy problem. If you partly understand, you may have to work very hard to get the answer.
If you don't understand, you'll have to guess true or false and won't be able to offer an explanation. (That's why it is important to always justify and explain when a problem asks you to. If you don't, it looks as if you don't understand and are just guessing.)

You don't need to substitute point A's messy coordinates into the inequality. While you can determine the right answer that way (and will be able to justify your answer), you will be doing lots of extra work. You should be able to determine true or false simply by looking at the picture. Since the point A is NOT part of the solution (i.e., point A is not in the shaded region), the coordinates of point A would make the inequality false.

Now that we are confident in working with inequalities, we can use them in applications. One application of systems of inequalities is known as linear programming. Despite the name, linear programming has nothing to do with computers! Linear programming takes many of the skills we have been developing and ties them together in one big problem. You saw this in LS-36, the Toy Factory problem and in LS-46, the Sandy Dandy Dune Buggies, Inc. problem. In each of these problems, you are presented with a lot of information. By carefully organizing the information mathematically, you can solve the problem.

Reread LS-46 and let's look at some parts of this problem. (Note: the complete solution is NOT presented here since many teachers like to use this as an assessment question. But, if you need assistance, these hints may help get you started.)

The first step is to write several inequalities to represent the information presented, with x defined as the number of Sand Crabs built in a week and y defined as the number of Surf Mobiles built in one week. The problem tells you that the company can get parts to produce as many as 15 Sand Crabs and 12 Surf Mobiles in one week. From this piece of information we can write $0 \le x \le 15$ and $0 \le y \le 12$. (You should know why I included greater than or equal to zero ($0 \le x$ and $0 \le y$) and not just less than the amount ($x \le 15$ and $y \le 12$). (Remember! The number of machines produced must be positive. We can't have -2 Surf Mobiles!) Next, each Sand Crab requires five manifold clamps, so two Sand Crabs would need ten manifold clamps, three Sand Crabs would need 15 manifold clamps, etc. And x Sand Crabs would require 5x manifold clamps. Similarly, since each Surf Mobile requires three manifold clamps, two Surf Mobiles need six clamps, three Surf Mobiles need nine clamps, etc. And y Surf Mobiles would need 3y manifold clamps. The maximum number of manifold clamps available each week in 81. This means we can write $5x + 3y \le 81$. (Note: it is "$\le$" because the company does not have to--and may not be able to--use all of the clamps each week.)

We can do a similar analysis on suspension joints. A Sand Crab needs two suspension joints each so x Sand Crab needs 2x suspension joints. Each Surf Mobile needs six suspension joints so y Surf Mobiles would need 6y of them. There are only 78 suspension joints available each week so our inequality is $2x + 6y \leq 78$.

Lastly we must use the constraints on the time available. It takes 20 hours to make each Sand Crab, so x Sand Crabs will take 20x hours. If we had y Surf Mobiles, at 30 hours each, it would require 30y hours to be assembled. The total hours to make both of these is $20x + 30y$. What are the constraints on the number of hours? Each employee can only work 37.5 hours per week, but there are 12 employees. Therefore, we can write
$20x + 30y \leq 12(37.5)$ or $20x + 30y \leq 450$.

So we have five inequalities occurring simultaneously: $0 \leq x \leq 15$, $0 \leq y \leq 12$, $5x + 3y \leq 81$, $2x + 6y \leq 78$ and $20x + 30y \leq 450$. Graph these on the same set of axes and find the region that is common to all of them (known as the feasibility region).

Next you must come up with a "profit line". The profit on each Sand Crab is \$500 so x Sand Crabs gives 500x profit. At \$1000 each, y Surf Mobiles would yield 1000y in profits. Therefore, the profit can be represented with the expression $P = 500x + 1000y$. We would like the profit to be as large as it can be given the constraints. Initially, it might help to choose any value for the profit P and graph the line you create. You can find the maximum profit by "sliding" the profit line up and down, making sure it fits within the feasibility region.

Once you find the maximum profit, you are not quite done yet! Now it is up to you to summarize and explain all of this in your report. Do a nice job.

The topics which follow linear programming cover systems, but systems of equations again, not inequalities. When solving two equations with two unknowns, it helps to know that each equation represents a line. The lines may or may not cross. If the lines cross, the point of intersection is the solution to <u>both</u> equations, that is, it satisfies both equations simultaneously. In three dimensions, we have an analogous situation. We can have a system of three equations with three unknowns representing three planes. These planes (imagine three sheets of paper) can intersect in a variety of ways and visualizing these intersections helps explain the possible solutions.

Three planes may not intersect at all (be parallel) as in Figure 1, may intersect in a line as in Figure 2, or intersect in a point as in Figure 3. Although it is probably the most difficult to picture in your mind's eye, it is this third case we are most interested in.

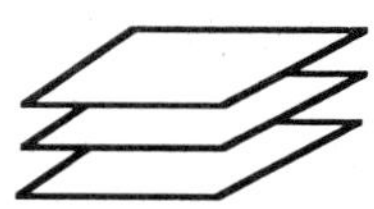

Figure 1

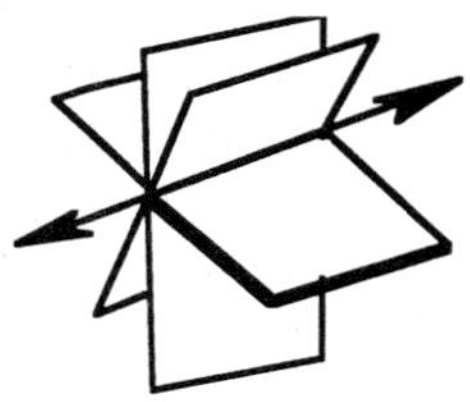

Figure 2

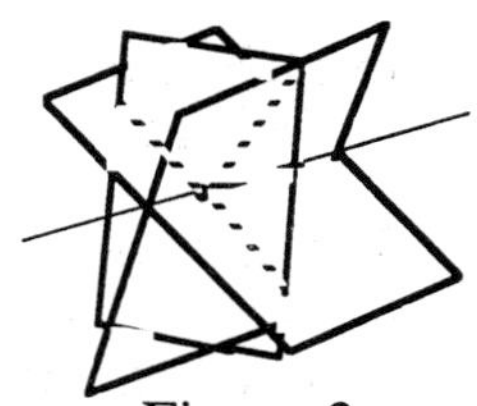

Figure 3

When solving three equations and three unknowns, we can use the same techniques as in solving two equations with two unknowns. Consider the following examples. Remember to try them yourself first!

6. Solve the following for x, y and/or z.

a) $$\begin{aligned} 5x + 4y &= -3 \\ 8x - 5y &= 18 \end{aligned}$$

b) $$\begin{aligned} 2x - y - z &= 7 \\ 3x + 5y + z &= -10 \\ 4x - 3y + 2z &= 4 \end{aligned}$$

Solution.

The equations in part (a) are familiar to us: two equations and two unknowns have been with us since Algebra 1! Here, we want to eliminate a variable so that we are left with <u>one</u> equation with <u>one</u> variable. To do this, I will multiply the top equation by five and the bottom equation by four. Be sure to multiply <u>everything</u> in each equation.

$$\begin{aligned} 5x + 4y &= -3 \quad \times 5 \rightarrow \quad 25x + 20y = -15 \\ 8x - 5y &= 18 \quad \times 4 \rightarrow \quad 32x - 20y = 72 \end{aligned}$$

Adding the new equations gives us $57x = 57$

which we can easily solve for x: $x = 1$. Next we substitute the x value back into either of the original equations to find y. Using the first equation,

$$5(1) + 4y = -3 \quad => \quad 4y = -8 \quad => \quad y = -2$$

Therefore the solution is $x = 1$, $y = -2$, or $(1, -2)$. (Note: this tells us that the two lines intersect in a point.)

We solve the three equations with three unknowns similarly. We still want to eliminate variables, there are just more of them to eliminate. To help in the explanation, I'm going to label the equations as shown at right. First, try adding equations Ⓐ and Ⓑ. What happens? The variable z should have been eliminated. (If not, check your work.) You should have been left with $5x + 4y = -3$. Now let's use a different pair of equations to eliminate z again, for instance, let's take two times equation Ⓐ and add the result to equation Ⓒ. Do this and be careful! You should be left with $8x - 5y = 18$.

Ⓐ $2x - y - z = 7$
Ⓑ $3x + 5y + z = -10$
Ⓒ $4x - 3y + 2z = 4$

So what have we accomplished? Well, by eliminating z we are now left with

$$\begin{aligned} 5x + 4y &= -3 \\ 8x - 5y &= 18 \end{aligned}$$

This should look familiar. We have reduced the problem to a simpler, similar problem. In fact, we reduced it to a problem we just solved in part (a)! (If you want the practice, solve for x and y again.) Solving these two equations will give us $x = 1$ and $y = -2$. Knowing x and y, we can use any of the three original equations with these values for x and y to find the value of z. Using equation Ⓐ we have.

$$2(1) - (-2) - z = 7 \quad => \quad 2 + 2 - z = 7 \quad => \quad -z = 3 \quad => \quad z = -3$$

So the solution is $x = 1$, $y = -2$, and $z = -3$.

By the way, the three equations represent three planes in space. Our solution tells us these three planes intersect at the point $(1, -2, -3)$.

Solving three equations with three unknowns can be tedious, but use what you have learned to solve two equations with two unknowns. Most common mishaps are careless mistakes, so be sure to show all your work. This way your teacher can look for understanding and you can look for mistakes. And, most importantly, just as with equations with one variable and equations with two variables, you can check your work by substituting the values back into each of the equations to be sure they check. To be correct however, please remember that the solution must work in <u>all</u> three equations.

Fortunately, we have the technology to help us avoid these tedious problems. By using a graphing calculator and understanding matrices, we can shrink these problems to a series of pokes at a few buttons. Initially, a matrix might seem strange and unuseful, but it simplifies the process of solving a system of equations. A matrix is a table of numbers and the numbers we will record in the matrix will often times be the coefficients of equations. We will also use matrices to hold information, acting as an organizer for us.

Consider the following example.

7. Han and Lea have been building jet roamers and pod racers. Each jet roamer requires two jet packs, one crystallic fuel tank and one turbo charger while each pod racer requires four jet packs, three crystallic fuel tanks and two turbo chargers. Han and Lea build these vehicles and sells them to the Empire, the Galactic Alliance and the Federation of Planets. To keep track of their supplies, products and profits, Han and Lea use matrices.

a) Use a matrix to represent the number of jet packs, crystallic fuel tanks and turbo chargers Han and Lea need to build the jet roamers and pod racers. Call the matrix M.

b) What are the dimensions of the matrix M?

c) How could you use matrix M to figure out the number of parts needed to make six jet roamers and six pod racers? Write this out, showing the answer.

Solution.

This problem contains several pieces of information. We will use a matrix to organize and hold the information for us. Let's set up a table with the column headings *jet packs, fuel tanks,* and *turbo chargers.* Each row of the table will represent the different vehicles.

	Jet packs	**Fuel tanks**	**Turbo chargers**
Jet Roamers	2	1	1
Pod Racers	4	3	2

If you can read and understand this table, then you can read and understand matrices. The '2' in the second row tells us the a pod racers needs two turbo chargers. If you need to know how many fuel tanks a jet roamer needs, find the number in the "fuel tank" column and the "jet roamer" row: 1. A matrix simply drops the headings of the rows and columns and adds square brackets.

$$M = \begin{bmatrix} 2 & 1 & 1 \\ 4 & 3 & 2 \end{bmatrix}$$

In part (b), we want the dimensions of this matrix. The dimensions are the number of rows and columns of the matrix. We write the number of rows first, thus the dimensions are 2 by 3; we usually write it as 2 x 3.

To figure out the number of parts needed to make six jet roamers and six pod racers, we multiply this matrix by 6: 6M. To multiply a matrix by six, we must multiply each term in the matrix by six.

$$6M = 6 \cdot \begin{bmatrix} 2 & 1 & 1 \\ 4 & 3 & 2 \end{bmatrix}$$
$$= \begin{bmatrix} 12 & 6 & 6 \\ 24 & 18 & 12 \end{bmatrix}$$

In the this new matrix we can easily read the number of parts we will need. For instance, to make six pod racers, Han and Lea will need 18 fuel tanks. Also, by adding the numbers in a column we see that they will need 36 jet packs total (12 + 24).

The problem continues.

8. Suppose the Gallactic Alliance orders 15 jet roamers and 12 pod racers. Write a new 1 x 2 matrix to represent this order.

a) Show how to use this matrix to determine the number of parts needed to fill the Gallactic Alliance's order.

b) Each jet roamer costs Han and Lea $4752 to make a jet roamer while a pod racer costs $3018 to make. How much will they spend to fill the Gallactic Alliance's order?

c) If Han and Lea sell each jet roamer for $8032 and each pod racer for $6215, how much profit will they make on the order?

Solution.
Just as before, we create a "table" which holds the information. The Gallactic Alliance's order can be written as

$$G = [15 \quad 12] .$$

I have labeled the matrix G to remind us it represents the Gallactic Alliance's order. To determine the number of parts needed to fill this order, as asked for in part (a), we will **multiply** matrix M by matrix G. Be sure you have read the box after LS-68 on matrix multiplication. When multiplying two matrices, there are two important points to remember. First, the two matrices must have "agreeable dimensions": the number of columns of the first matrix must equal the number of rows of the second matrix. Thus, we can multiply a matrix with dimensions p x q by a q x r matrix. (Notice how the two middle numbers are the same: q. That is the key. The result will be a p x r matrix.) Since our matrix G is a 1 x 2 matrix and M is a 2 x3 matrix, we can multiply these two matrices. The result will be a 1 x 3 matrix.

$$GM = [15 \quad 12] \cdot \begin{bmatrix} 2 & 1 & 1 \\ 4 & 3 & 2 \end{bmatrix}$$

For some reason, what helps me remember how to do matrix multiplication correctly is to use both hands. I walk my left hand through the rows on the left matrix while my right hand walks through the columns on the right matrix.

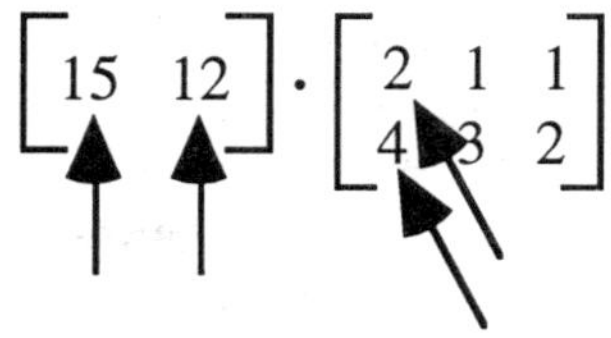

We multiply the first ones and add to the product of the second ones: $15 \cdot 2 + 12 \cdot 4$. The answer, 78, is the first element in the 1 x 3 matrix. We do this same process of multiplying and adding again, this time using the second column.

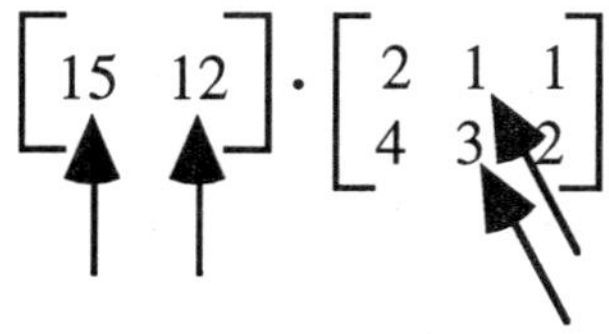

Here, $15 \cdot 1 + 12 \cdot 3 = 51$ is the number in the second column. Lastly, we use the third column to get $15 \cdot 1 + 12 \cdot 2 = 39$. Our matrix looks like:

$$GM = [15 \quad 12] \bullet \begin{bmatrix} 2 & 1 & 1 \\ 4 & 3 & 2 \end{bmatrix} = [78 \quad 51 \quad 39]$$

This matrix tells us it will take 78 jet packs, 51 fuel tanks and 39 turbo chargers to complete the order.

In part (b) we now have to worry about the cost to make these. Again, let's make a matrix that represents how much it costs to make these vehicles. I have named the matrix C.

$$C = \begin{bmatrix} 4752 \\ 3018 \end{bmatrix}$$

If we multiply GC we will determine the total cost for making these vehicles for the Gallactic Alliance.

$$GC = [15 \quad 12] \bullet \begin{bmatrix} 4752 \\ 3018 \end{bmatrix} = [107496]$$

Therefore, to fill the Gallactic Alliance's order, Han and Lea will spend \$107,496. How much will they make? Part (c) asks for the profit after supplying the profit for each one. Again we will multiply by a matrix. The matrix we will use I'll call matrix P for profit.

$$GP = [15 \quad 12] \bullet \begin{bmatrix} 8032 \\ 6215 \end{bmatrix} = [195060]$$

If they make \$195,060 when they deliver the order and it cost them \$107,496 to fill the order, Han and Lea will make \$195,060 - \$107,496 = \$87,564.

A legitimate question is: how do I know whether the matrix should be a row matrix (1 x 2 say) or a column matrix (2 x 1)? Another important question: how do I know whether to multiply GM or MG? The answer I'll give might be unsatisfying but it is correct: you do what you need to do to make it work. In the example above, if I had made matrix P a 1 x 2 matrix, I couldn't have multiplied GP (G is a 1 x 2 and a 1 x 2 and 1 x 2 don't agree dimensionally). In the earlier problem, G is a 1 x 2 matrix and M is a 2 x 3 matrix. For the dimensions to agree we must multiply GM. Don't be afraid to try something, and if it doesn't work, try something else. Switch the order or dimensions. Remember, after all, matrices are just tables of information. You can set up your table any way you want.

As mentioned earlier, matrices can also be used to solve systems of equations. When using matrices in conjunction with a graphing calculator, even solving a system of three equations with three unknowns is relatively simple. The key is entering the appropriate matrix and performing the appropriate operations. We'll look at the basics of this here and leave it to you and/or your teacher to learn the specifics for your calculator and your class. (Note: some teachers may not want you to use a graphing calculator to solve a system of equations. Check with your teacher before doing so.)

When multiplying two matrices, we perform some products and sums. Thus, in the product

$$\begin{bmatrix} 8 & -1 & 4 \\ 3 & 2 & 2 \\ -5 & 0 & 6 \end{bmatrix} \bullet \begin{bmatrix} x \\ y \\ z \end{bmatrix}$$

the first element in the resulting 3 x 1 matrix is 8x - 1y + 4z. The second row is 3x + 2y + 2z and the third is -5x + 0y + 6z. (Verify this yourself!) It should be written as

$$\begin{bmatrix} 8x - 1y + 4z \\ 3x + 2y + 2z \\ -5x + 0y + 6z \end{bmatrix}$$

This <u>almost</u> looks like a system of three equations with three unknowns. What is missing? The "equals to" part!

So now suppose we had a system of three equations with three unknowns:

$$\begin{aligned} 8x - y + 4z &= 15 \\ 3x + 2y + 2z &= -2 \\ -5x + 0y + 6z &= -16 \end{aligned}$$

We rewrite this as the matrix equation

$$\begin{bmatrix} 8 & -1 & 4 \\ 3 & 2 & 2 \\ -5 & 0 & 6 \end{bmatrix} \bullet \begin{bmatrix} x \\ y \\ z \end{bmatrix} = \begin{bmatrix} 15 \\ -2 \\ -16 \end{bmatrix}$$

If we let

$$A = \begin{bmatrix} 8 & -1 & 4 \\ 3 & 2 & 2 \\ -5 & 0 & 6 \end{bmatrix}, \ \mathbf{x} = \begin{bmatrix} x \\ y \\ z \end{bmatrix} \text{ and } B = \begin{bmatrix} 15 \\ -2 \\ -16 \end{bmatrix}$$

we can write this system as the matrix equation A**x** = B. This is easily solved for **x**.

$$A\mathbf{x} = B$$
$$A^{-1} \bullet A\mathbf{x} = A^{-1} \bullet B$$
$$\mathbf{x} = A^{-1} \bullet B$$

This is where your graphing calculator comes in. Once you enter matrix A, a 3 x 3 matrix, and matrix B, a 3 x 1 matrix, you can find **x** by computing $A^{-1} \cdot B$. Try it! You should (quickly) get a 3 x 1 matrix

$$\begin{bmatrix} 2 \\ -3 \\ -1 \end{bmatrix}$$

which means x = 2, y = -3 and z = -1.

But why do we need to know how to solve three equations with three unknowns? It may be surprising to you but solving three equations and three unknowns allows us to find the equations of parabolas given three points. Then we can use this technique to determine mathematical models which can be represented by a parabola. Consider the following problem.

9. Pizza Palace sells three sizes of combination pizzas.

Small (8" diameter)	**\$8.50**
Medium (10" diameter)	**\$11.50**
Large (13" diameter)	**\$17.50**

Assume that the price of the pizza can be modeled with a quadratic function, with the price dependent on the diameter of the pizza.

a) What three data points can we write from the above information?

b) Write the equation representing the data points.

c) Use your equation to predict the cost of a combination pizza with an 18" diameter.

d) What if a pizza cost \$50.00. How big is it?

Solution.

The three data points are (8, 8.50), (10, 11.50), and (13, 17.50). The variable x represents the diameter, in inches, of the pizza and y is the cost, in dollars, of the pizza. Something you should be thinking about is what are the domain and range for this situation? From the information given, all we know is that the domain and range should be non-negative, i.e. the domain is $x \geq 0$ and the range is $y \geq 0$.

In part (b), we want to find a quadratic function (a parabola) which passes through these three points. The general equation for a parabola is $y = ax^2 + bx + c$. We want to find a, b, and c so we have an equation that our three data points will satisfy. Substituting the coordinates for the first point for x and y we have:

$$8.50 = a(8)^2 + b(8) + c$$

Ⓐ $$8.50 = 64a + 8b + c$$

I've labeled this equation Ⓐ. We do the same thing for the second and third point to give equations Ⓑ and Ⓒ:

Ⓑ $$11.50 = 100a + 10b + c$$

Ⓒ $$17.50 = 169a + 13b + c$$

Now we must solve this system of three equations and three unknowns to find the equation of the parabola. The unknowns here are a, b, and c. If your teacher approves, rewrite this system as a matrix equation and use your graphing calculator to solve. If not, you can solve this by hand. I show the "by hand" solution here. Either way you should get the same answer.

Luckily with these problems, it is usually easy to eliminate c. To see this, subtract equation Ⓐ from equation Ⓑ and subtract equation Ⓑ from equation Ⓒ.

$$\text{Ⓑ - Ⓐ} \quad \rightarrow \quad 3 = 36a + 2b$$
$$\text{Ⓒ - Ⓑ} \quad \rightarrow \quad 6 = 69a + 3b$$

Now we are back to the simpler case of two equations with two unknowns. You should be pretty good at solving these by now! I will solve this by multiplying the top equation by -3 and the bottom equation by 2 and adding the two equations.

$$\begin{aligned} -9 &= -108a - 6b \\ \underline{12} &= \underline{138a + 6b} \\ 3 &= 30a \\ a &= \frac{1}{10} \end{aligned}$$

Substituting back into the first equation,

$$\begin{aligned} 3 &= 36\left(\frac{1}{10}\right) + 2b \\ 3 &= \frac{36}{10} + 2b \\ 3 &= 3.6 + 2b \\ -.6 &= 2b \\ b &= -.3 = -\frac{3}{10} \end{aligned}$$

We still have to find c. We can use any of the three original equations but all the "tenths" tell me it will be easiest to use equation Ⓑ.

$$\begin{aligned} 11.50 &= 100\left(\frac{1}{10}\right) + 10\left(\frac{-3}{10}\right) + c \\ 11.50 &= 10 - 3 + c \\ c &= 4.50 \end{aligned}$$

Don't lose sight of what we are doing! Finding a, b, and c is only part of the problem. Knowing a, b, and c lets us write the equation of a parabola that passes through the three data points. The equation is $y = \frac{1}{10}x^2 - \frac{3}{10}x + 4.50$.

Now that we have an equation, we can answer the question in part (c). To predict the cost of an 18 inch combination pizza, we can substitute x = 18 into our equation.

$$\begin{aligned} y &= \frac{1}{10}(18)^2 - \frac{3}{10}(18) + 4.50 \\ y &= \frac{324}{10} - \frac{54}{10} + 4.50 \\ y &= \$31.50 \end{aligned}$$

Therefore, a combination pizza with an 18 inch diameter would cost $31.50 according to the model.

Now let's do part (d). What if the pizza cost \$50? Let's set the equation equal to 50 and solve for x to find the diameter.

$$\frac{1}{10}x^2 - \frac{3}{10}x + 4.50 = 50$$

Multiply everything by 10 -- fraction busters!

$$x^2 - 3x + 45 = 500$$
$$x^2 - 3x - 455 = 0$$

Now use the Quadratic Formula!

$$x = \frac{3 \pm \sqrt{9 - 4(-455)}}{2}$$
$$x = \frac{3 \pm \sqrt{1829}}{2}$$
$$x \approx \frac{3 \pm 42.66}{2}$$

We only concern ourselves with positive answers since a pizza can't have a negative length diameter. Therefore, $x \approx 22.83$ inches.

Problems like this last one are BIG problems and it is easy to get lost and forget where you are going. Remember your subproblems! If it would help, begin the problem by listing an "outline" of all the subproblems you must do:

1. Write three equations with three unknowns.
2. Solve the three equations with three unknowns.
3. Use the solution to write the equation of a parabola.

Earlier I told you it helps to look for simpler, similar cases of problems. That is particularly true in this unit. Solving three equations with three unknowns is done in the same way as two equations with two unknowns. Also watch for small, careless errors. They can lead to redoing quite a bit of work. Be sure to check your work whenever possible. Not only could you catch a possible mistake, it will also show your teacher your understanding and your concern for doing a complete, successful job.

UNIT 5
MORE TO TRY

Solve each of the following systems for x, y, and/or z. Then explain what the answer tells you about the graphs of the equations. Be careful and check your work.

1. $8x + 3y = 54$
   $5x + 4y = 38$

2. $7x + 4y = 68$
   $5x + 2y = 46$

3. $3x + 4y = 29$
   $7x + 6y = 41$

4. $2x + 5y = 43$
   $-x + 3y = 71$

5. $15x + 10y = 21$
   $18x + 12y = 35$

6. $3x - 2y + z = 3$
   $5x + y + 2z = 8$
   $-3x - y + 3z = -22$

7. $-4x - 6y + 5z = 21$
   $3y + 4y - 2z = -15$
   $-7x - 5y + 3z = 15$

8. $3x + 4z = 19$
   $3y + 2z = 8$
   $4x - 5y = 7$

9. $9x + 6y - 12z = 14$
   $3x + 2y - 4z = -11$
   $x + y + z = 1$

10. $\frac{x}{3} + \frac{y}{4} + \frac{z}{2} = 24$
    $\frac{x}{2} + \frac{y}{3} + \frac{z}{4} = 29$
    $\frac{x}{4} + \frac{y}{2} + \frac{z}{3} = 25$

Graph the following systems of inequalities. Darken in the solution (the region satisfying all inequalities).

11. $y < \frac{1}{2}x + 6$
    $y > -\frac{1}{2}x + 6$
    $x < 12$

12. $x + y < 10$
    $x + y > 4$
    $y < 2x$
    $y > 0$

13. $y < 3x + 4$
    $y > -\frac{1}{4}x + 8$
    $y > -\frac{1}{3}x + 4$
    $y > 5x - 6$

14. $3x + 4y < 12$
    $y > (x + 1)^2 - 4$

15. $y < -\frac{3}{4}(x - 1)^2 + 6$
    $y > x - 7$

16. $y < (x + 2)^3$
    $y > x^2 + 3x$

Write the system of inequalities that when graphed, will produce these regions.

17.

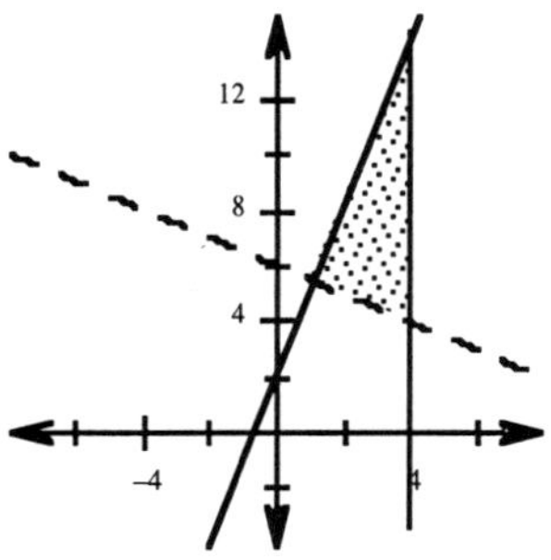

18.

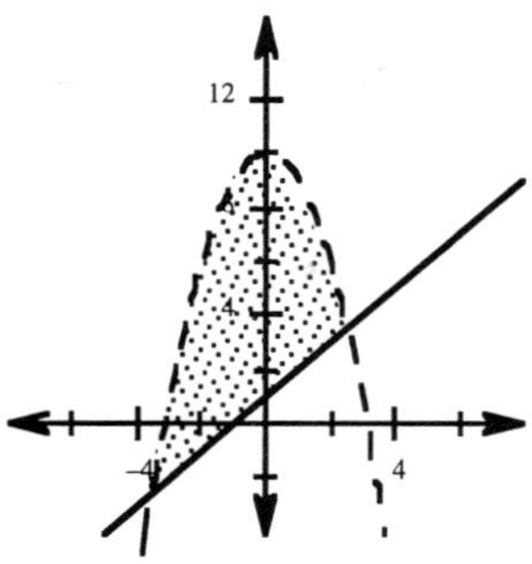

19. Find the equation of the parabola passing through these three points: (-2, -32), (0, -10), and (2, -12).

20. While reading a recent study done on various aged people, you notice a trend. The study counted the number of times in a 24 hour period that the people misunderstood or misinterpreted a statement, comment or question. The study offers these numbers.

Age (years)	**Misunderstandings or Misinterpretations**
20	44
30	28
40	20

You believe that the number of misunderstandings should reach a minimum at some age then go up again for very old people. Therefore, you assume that a quadratic function will best model this data. Find the equation which best fits this data. Use your equation to predict how many times an 80 year person will misunderstand or misinterpret. What about a one-year old? Who understands the most?

21. In archery, the arrow appears to travel in a straight line when it is released. But, the arrow will actually travel upward slightly before curving back down toward the earth. For a particular archer, the arrow starts at 5.4 feet above the ground. After 0.3 seconds, the arrow is 5.5 feet above the ground. The arrow hits the target after a total of 2 seconds at a height of 5 feet above the ground. Find the particular equation which models this data.

22. Davis and Tess are considering opening their own business. They wish to make and sell little alien dolls they call "Hauteans" and "Zotions." Each Hautean sells for $1.00 while each Zotion sells for $1.50. They can make up to 20 Hauteans and up to 40 Zotions, but no more than 50 aliens total. When Davis and Tess go to city hall to get a business license, they find there are a few more restrictions on their production. The number of Zotions (the more expensive item) can be no more than three times the number of Hauteans (the cheaper item). How many of each alien should Davis and Tess make to maximize their profit? What will the profit be?

23. Reread problems 7 and 8 presented in this section. Suppose the Federation of Planets wishes to purchase nine jet roamers and 21 pod racers. Using matrices, determine the number of parts Han and Lea will need and what their profit will be.

SAT PREP

1. If $\frac{x + 4}{12} = \frac{4}{3}$, then x equals

   a) 3 b) 6 c) 8 d) 10 e) 12

2. What is the least of three consecutive integers whose sum is 21?

   a) 5 b) 6 c) 7 d) 8 e) 9

3. Juanita has stocks, bonds, and t-bills for investments. The number of t-bills she has is one more than the number of stocks, and the number of bonds is three times the number of t-bills. Which of the following could be the total number of investments?

   a) 16 b) 17 c) 18 d) 19 e) 20

4. Through how many degrees would the minute hand of a clock turn from 5:20 PM to 5:35 PM the same day?

   a) 15° b) 30° c) 45° d) 60° e) 90°

5. The length of a rectangle is six times its width. If the perimeter of the rectangle is 56, what is the width of the rectangle?

   a) 4 b) 7 c) 8.5 d) 18 e) 24

6. If $m > 1$ and $m^n m^5 = m^{15}$, then what does n equal?

7. In the triangle at right, what is the value of a + b + c + d?

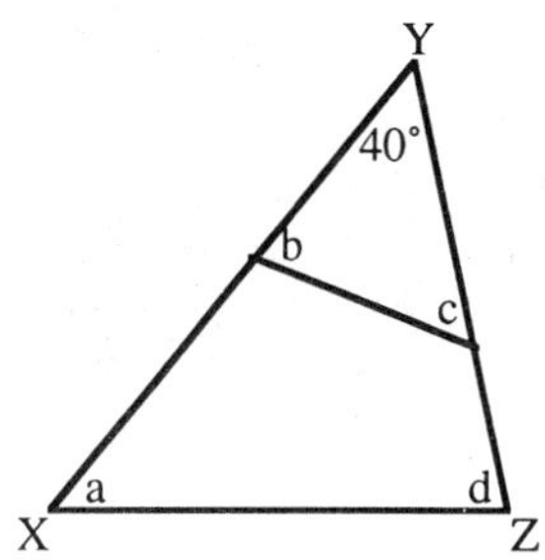

8. If x and y are positive integers, $x + y < 12$, and $x > 4$, what is the greatest possible value for $x - y$?

9. If $(2x^2 + 5x + 3)(2x + 4) = ax^3 + bx^2 + cx + d$ for all values of x what does c equal?

10. Four lines intersect in one point creating eight congruent adjacent angles. What is the measure of one of these angles?

UNIT 6

The Case of the Cooling Corpse

LOGARITHMS AND OTHER

UNIT 6
LOGARITHMS AND OTHER INVERSES

As you work through this unit (and the course for that matter) you will frequently come across equations that you will find difficult or tedious to solve. The only way to solve some of these may be by guess and check. You probably did this with the equation $\sin x = 0.75$ or $1.04^x = 2$. In this unit we will find ways to solve these equations precisely without guess and check.

You can easily solve some equations, such as $3x + 2 = 8$, by "undoing" the equation. Some equations however, cannot be undone without the creation of a new function. These "undoing" functions are called **inverse** functions and a function and its inverse have some interesting relationships. To see this let's start with a simple example.

1. Find the inverse of $f(x) = \frac{1}{5}x + 3$. Graph $f(x)$ and its inverse on the same set of axes.

Solution.
We want to "undo" f(x). Think of "unwrapping" the coefficients and constants in this equation, until we reach what is on the inside: x. There are several ways to do this and you should use the method you can use correctly consistently. One way is to think of inverses as an interchange function. You saw this in CC-29. First, I will rewrite the equation replacing f(x) with y:

$$y = \frac{1}{5}x + 3$$

Then interchange the x and y:

$$x = \frac{1}{5}y + 3$$

and solve for y.

$$\begin{aligned} x &= \frac{1}{5}y + 3 \\ x - 3 &= \frac{1}{5}y \\ 5(x - 3) &= 5\left(\frac{1}{5}y\right) \\ y &= 5x - 15 \end{aligned}$$

This is the inverse of f(x); let's call it g(x). The graph of both of these functions is shown at right.

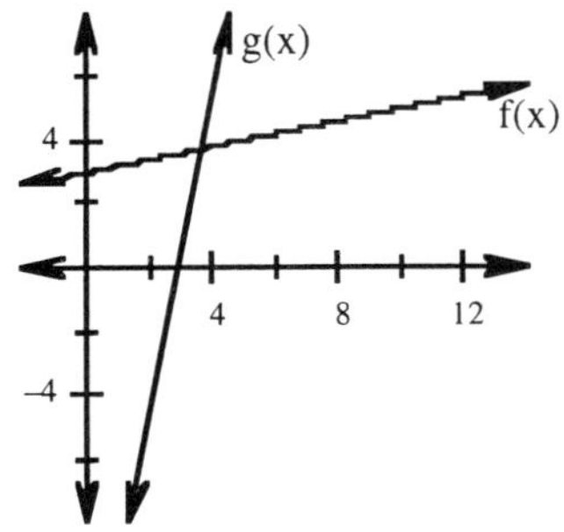

This method of interchanging the x and y variables and then solving for y is a good algorithm to know. Be sure you have an example (with helpful notes) in your tool kit. Try it out on the next one. Remember! Do it on your own first!

2. Find the inverse of $f(x) = x^2 + 6$. Graph both f(x) and its inverse on the same set of axes.

Solution.
As before, we start by rewriting the function with y, then interchanging the two variables, and finally solving for y:

$$\begin{aligned} y &= x^2 + 6 \\ x &= y^2 + 6 \\ x - 6 &= y^2 \\ y &= \pm\sqrt{x - 6} \end{aligned}$$

Did you remember to write ±? When taking the square root, there are two possible answers: a positive one and a negative one. Just remember that if $x^2 = 25$, then x = 5 **or** -5.

To graph the function and its inverse, it is as if we are graphing three functions: $f(x) = x^2 + 6$, $y_1 = \sqrt{x - 6}$ and $y_2 = -\sqrt{x - 6}$.

With both functions we have looked at, the graph of the inverse is actually the reflection of the original function across the line y = x. You saw this in CC-16 and CC-17 and it is clear in the graph at right. In the previous example, sketching in the graph y = x might make it clearer. Knowing that the inverse is the reflection of the function across the line y = x lets you quickly sketch the inverse even before you know the equation of the inverse. CC-18 describes this procedure.

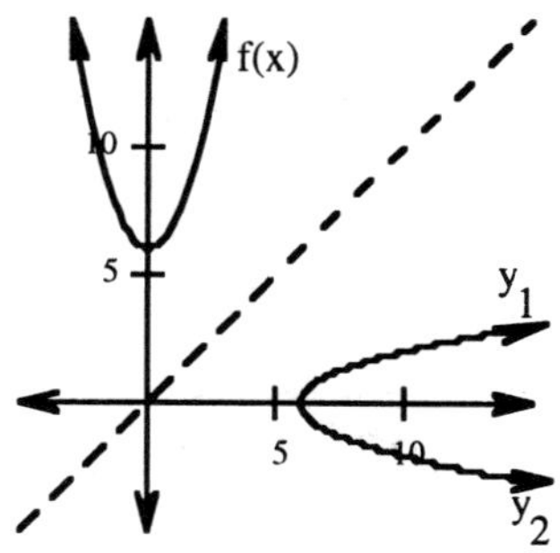

So inverses are undoing or interchange functions. Let's go back to that troublesome equation $1.04^x = 2$. We want to be able to solve this equation (and others like it) for x accurately rather than guessing and checking. How do we undo this equation? To undo (i.e. solve) this equation we need a new function. This happens in mathematics. Mathematicians invent new mathematics to allow them to do more mathematics. In this case, they invented a new function called a logarithm and it "undoes" exponential functions. To help explain the relationship, let's look at a table of values for the exponential function $f(x) = 3^x$.

x	-3	-2	-1	0	1	2	3
y	$\frac{1}{27}$	$\frac{1}{9}$	$\frac{1}{3}$	1	3	9	27

The inverse of this exponential function will have a table of values similar to this, just with x and y interchanged. Here is the table for the inverse:

x	$\frac{1}{27}$	$\frac{1}{9}$	$\frac{1}{3}$	1	3	9	27
y	-3	-2	-1	0	1	2	3

We call this new function **logarithm**, base 3. We write $y = \log_3 x$ and read it "y equals log, base 3 of x." From this table we see that $\log_3 27 = 3$ and $\log_3 9 = 2$.

In this example, the logarithm is in base 3 but logs can be in any positive base just as an exponential function can be in any base. Calculating numeric values of different logarithmic expressions can be perplexing. Your calculator only finds values of logarithms in base 10. Try log 100 on your calculator now. You should get an answer of 2 because (and here is the part you need to understand) $10^2 = 100$. A logarithm is really just an exponent. By remembering this fact, we can calculate the value of an expression such as $\log_4 64$ even though our calculator does not work in base 4. This expression, $\log_4 64$, is the exponent we raise 4 to, to get 64. What number can we raise 4 to to get 64? Three. Therefore, $\log_4 64 = 3$.

For each logarithmic equation there corresponds an exponential equation and vice versa. Knowing this correspondence will help you understand logarithms. For instance, for the exponential equation $3^4 = 81$, the corresponding logarithmic equation is $4 = \log_3 81$. If we start with $\log_4 2 = \frac{1}{2}$ then the corresponding exponential equation is $4^{1/2} = 2$. Note: When no base is written, as in log 100, the base is 10. This is just a convention you must memorize and/or make note of in your tool kit.

With practice you will be able to look at equations such as $\log_5 25 = x$ and solve them mentally just as you can solve $5^x = 25$ mentally. Try a few now.

3. Solve for x.

a) $\log_7 49 = x$ **b)** $\log_3 81 = x$ **c)** $\log_x 16 = 2$

d) $\log_x 64 = 3$ **e)** $\log_6 x = 3$ **f)** $\log_2 x = 5$

Solution.
Some of you may already be at the point where you can mentally solve these problems. That's great! If you check your answers as well, you are set. Sometimes just stating the meaning of the equation allows you to solve the equation mentally. I've done that on some of these. I will also rewrite it as an exponential equation. If you are having trouble solving this type of equation, be sure to rewrite it as an exponential equation. If you can do nothing else, this will show your teacher that you see the relationship.

In part (a), x is the number we raise 7 (the base) to, to get 49. As an exponential equation, $7^x = 49$. Now you should be able to solve this mentally: $x = 2$. (b): x is the number we raise 3 to, to get 81. $3^x = 81$, $x = 4$. (c): 2 is the number we raise x to, to get 16. $x^2 = 16$, $x = 4$. (The base of a logarithm is positive.) (d): $x^3 = 64$, $x = 4$. (e): $6^3 = x$, $x = 216$. (f): $2^5 = x$, $x = 32$.

You have solved exponential equations before by making the bases the same. Recall that to solve $4^x = 8^{x+2}$ we had to change the base to two.

$$\left(2^2\right)^x = \left(2^3\right)^{x+2}$$

$$2^{2x} = 2^{3x+6}$$

so $2x = 3x + 6$ and $x = -6$.

Remember the equation $1.04^x = 2$? How can we make the bases the same in this problem? Logarithms will help. We will change everything to base 10 because that is the base that our calculator uses. First we want to write 2 as a power of 10 or solve $2 = 10^x$. If we write this in logarithm form we are done: x = log(2). (Remember that this means base 10!) So the right side of our equation can be replaced with $10^{\log(2)}$. Writing 1.04 as a power of 10 means solving $1.04 = 10^x$. Again writing this in logarithm form we get x = log(1.04) so 1.04 = $10^{\log(1.04)}$ which can be substituted into the left side. Our origional equation now becomes:

$$10^{\log(1.04)x} = 10^{\log(2)}$$

so since the bases are the same,

$$\log(1.04)x = \log(2).$$

But log(1.04) and log(2) are just numbers so we solve the equation by dividing both sides by log(1.04).

$$\frac{\log(1.04)x}{\log(1.04)} = \frac{\log(2)}{\log(1.04)}$$

$$x = \frac{\log(2)}{\log(1.04)}$$

Use the calculator to find a numeric approximation for this.

$$x \approx 17.67298769 \ldots$$

4. Use logs to solve each of these accurate to five decimal places.

a) $7^x = 10$ b) $5.6^x = 3$ c) $1.0275^x = 100$

Solution.

a) $7^x = 10$

$10^{\log(7)x} = 10^{\log(10)}$

$\log(7)x = \log(10)$

$x = \frac{\log(10)}{\log(7)}$

$x \approx 1.18329$

b) $5.6^x = 3$

$10^{\log(5.6)x} = 10^{\log(3)}$

$\log(5.6)x = \log(3)$

$x = \frac{\log(3)}{\log(5.6)}$

$x \approx .63770$

c) $1.0275^x = 100$

$10^{\log(1.0275)x} = 10^{\log(100)}$

$\log (1.0275)x = \log 100$

$x = \frac{\log(100)}{\log(1.0275)}$

$x \approx 169.75291$

Do you see a pattern here? To solve $a^x = b$; $x = \frac{\log(b)}{\log (a)}$.

Now that we can solve these exponential equations accurately, we can solve some applications. A typical application of this sort involves money. Try this on your own first.

5. Karen has \$5000 to invest for her children's college education. Her local bank offers a 7.5% money market, which compounds the interest quarterly. If she wants to have at least \$12,000 for college, for how long must she keep the money in the account?

Solution.

From our earlier work, we know that money in an interest earning account grows exponentially according to the equation $y = am^x$ where a is the initial amount invested, m is the interest rate, x is the length of time the money is invested, and y is the ending balance. In this problem we have

$$12000 = 5000(1.01875)^x$$

First, do you know where the 1.01875 came from? The interest rate is 7.5% compounded quarterly so we divide .075 by four to find the rate at which interest is calculated at each quarter. Also, rather than calculating the interest and adding it to the initial amount each time (5000 + 5000(.01875)) we shorten this with mathematics to 5000(1 + .01875) or 5000(1.01875). (If you have forgotten this, look back to our review of Unit 3.)

Until recently if you wanted to solve this equation you would have to use guess and check or graph both sides of the equation and find where they cross. Now thanks to logarithms, we can quickly and easily solve this.

$$12000 = 5000(1.01875)^x$$

$$\frac{12000}{5000} = \frac{5000(1.01875)^x}{5000}$$

$$2.4 = 1.01875^x$$

$$x = \frac{\log 2.4}{\log 1.10875}$$

$$x \approx 47.128$$

Use the pattern developed in problem four to solve for x.

Now comes the part of the problem where you must think. What does 47.128 mean? Remember that in the equation $y = am^x$, x is the time the money will be in invested, and in this problem, it is the number of quarters. But can Karen invest the money for 47.128 quarters? Banks usually calculate the interest at the end of the quarter. At the end of the 47th quarter, Karen will have \$11971.49 (check this yourself!), not the \$12000 she wants. After 48 complete quarters, the amount is \$12,195.96. Thus, leaving the money in the money market account for 48 quarters -- or 12 years -- will give Karen the amount she wants.

Now that we have logarithms we can also find a specific equation to fit exponential data. You did this in several problems in this unit, and all are done similarly. There are two variations presented here: exponential data with the x-axis as an asymptote and data with a different horizontal asymptote. Before trying this next example, you may wish to reread problems CC-127, CC-137, CC-138 and CC-139 and their solutions.

---

6. A car costs \$12,000 new. After five years its resale value (known as its "blue book" value for the book that lists used car values) is \$7,500. Write an equation which will model this data and use it to find when the car is worth \$1,000.

Solution.

As we have seen (and this also makes sense) the value of the car decreases, but it will never be worth nothing (\$0). This data will best be modeled by an exponential equation such as $y = km^x$. This type of equation has a horizontal asymptote at the x axis and so does this data. (I said the value will never be \$0, but it will get close!) The first thing to figure out is: what are the data points in this problem? They are (0, 12000) and (5,7500). Substituting these values for x and y in the equation gives

$$12000 = km^0$$

$$7500 = km^5$$

The first equation simplifies nicely: since $m^0 = 1$, $k = 12000$. Let's use this piece of information in the second equation and solve for m.

$$\begin{aligned} 7500 &= 12000m^5 \\ \frac{7500}{12000} &= \frac{12000m^5}{12000} \\ .625 &= m^5 \\ (.625)^{1/5} &= (m^5)^{1/5} \\ m &\approx .91028 \end{aligned}$$

Now that we have k and m, we can write the equation: $y = 12000(.91028)^x$. Let's use this to find when the car is \$1000. Letting $y = 1000$, we have

$$1000 = 12000(.91028)^x$$

We will use logs to solve for x. (You did try solving this on your own first, right?)

$$\begin{aligned} \frac{1000}{12000} &= \frac{12000(.91028)^x}{12000} \\ \frac{1}{12} &= (.91028)^x \\ x &= \frac{\log\left(\frac{1}{12}\right)}{\log(.91028)} \\ x &\approx 26.43 \end{aligned}$$

This tells us that in approximately 26.43 years, the car will be worth about \$1000.

Now let's look at another example in which the horizontal asymptote is not the x axis. We will use the data from a problem we looked at in our review of Unit 3, but we will now find the equation -- something we did not have the skills to do back then.

7. On a cold wintry day, as Karen made herself a cup of cocoa, she decided to conduct a mini science experiment as well. She took a mug and filled it with cocoa, placed a thermometer in the cocoa, and left it sitting next to her as she worked. Every couple of minutes or so, she wrote down the time and the reading on the thermometer. The data is shown below.

Time since 1st reading	**0**	**1**	**5**	**7**	**10**	**12**	**15**	**27**
Temp (°F)	**142°**	**128.7°**	**96.4°**	**88.01°**	**80.5°**	**77.6°**	**75.0°**	**72.2°**

a) Plot the points. Describe any trends you see.

b) What sort of function would best model/fit this data? Explain. What does the general equation of this sort of function look like?

c) Find the equation which fits this data.

Solution.

We solved parts (a) and (b) in our review of Unit 3. Part (c) is new. What we found in part (a) is that the points appeared to lie on a decreasing exponential curve, leveling off at about 72. This told us that the data had a horizontal asymptote at y = 72°. We will use this fact along with only two data points, to find the equation.

The general equation of an exponential function looks like $y = km^x$, **if** the horizontal asymptote is the x axis. Here the horizontal asymptote is the line y = 72. This will change the general equation. If we want the exponential function to have a horizontal asymptote at y = c, the general equation becomes $y = km^x + c$. So for our cocoa problem, the equation will look like $y = km^x + 72$. We just need to find out what k and m are. We will do this similarly to what we did in the previous example. We can use any of the data points in the table. I always like to use points that will make my work easier, so I will use the first two points: (0, 142) and (1, 128.7), in the hope of keeping the numbers small. Substituting these values into our equation for x and y we have:

$$\begin{aligned} 142 &= km^0 + 72 \text{ , and} \\ 128.7 &= km^1 + 72 \end{aligned}$$

The first equation simplifies nicely. Let's work with that one first.

$$\begin{aligned} 142 &= km^0 + 72 \\ 142 &= k + 72 \\ k &= 70 \end{aligned}$$

Use this value of k in the second equation above to find m:

$$\begin{aligned} 128.7 &= 70m + 72 \\ 56.7 &= 70m \\ m &= \frac{56.7}{70} \\ m &\approx .81 \end{aligned}$$

Therefore, the equation appears to be $y = 70(.81)^x + 72$. To check if this seems reasonable, I will use another data point, (12, 77.6), and substitute the values in for x and y.

$$\begin{aligned} 77.6 &\stackrel{?}{=} 70(.81)^{12} + 72 \\ 77.6 &\stackrel{?}{=} 5.584 + 72 \\ 77.6 &\approx 77.584 \end{aligned}$$

This is reasonably close, so I believe it is correct. Note: with many applications, numbers may not match exactly. There can be round-off errors which effect the results. If we doubted this we could use another data point to check again. Do this if you wish to have more practice.

Unit 6 provides us with many tools and many new ideas. Logarithms are different, but very useful. What you need to do is be prepared to practice, practice, practice! If you come across a problem you have trouble with, try to relate it back to a simpler, similar problem that you can do. (Of course, that is a good rule to follow in any unit!) Have your tool kit complete with examples of how to solve different equations, and be sure to include several that require logarithms. But also include examples of problems that don't use logs to solve. Logarithms are useful, but they don't work on every problem. Know when to use them, as well as how.

UNIT 6
MORE TO TRY

Find the inverse of each of the following.

1. $f(x) = 8(x - 13)$
2. $y = -\frac{3}{4}x + 6$
3. $y = \frac{5(x + 2)}{3}$
4. $f(x) = x^2 + 6$
5. $f(x) = \frac{3}{x} + 6$
6. $g(x) = \frac{5}{x}$
7. $g(x) = (x + 1)^2 - 3$
8. $y = (x + 2)^3$

Sketch a graph of the <u>inverse</u> of each of these functions.

9. $y = \frac{x}{6} + 2$
10. $f(x) = 2x^2 - 1$
11.

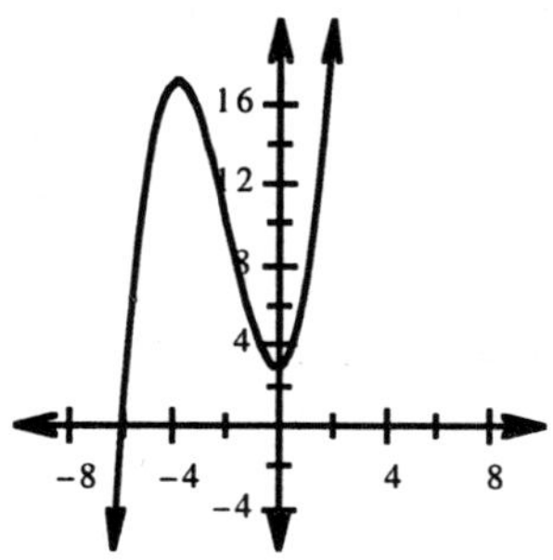

Rewrite each logarithmic equation as an exponential equation and vice versa.

12. $y = \log_4 x$
13. $3 = \log_2 x$
14. $x = \log_5 30$
15. $4^x = 80$
16. $(.5)^x = 64$

Solve for x.

17. $(2.3)^x = 7$
18. $12^x = 6$
19. $\log_7 49 = x$
20. $\log_3 x = 4$
21. $5(3.14)^x = 18$
22. $7x^8 = 294$
23. $\log_x 100 = 4$
24. $\log_5 45 = x$

The graph of an exponential function passes through the points (0,7) and (3, 189). Find the equation of the function if

25. the horizontal asymptote is the x axis.
26. the horizontal asymptote is the line $y = 8$.

27. At 1:00 am, a homicide detective found the reading of a corpse's body temperature to be 88°F. One hour later, the body temperature is 83°. If the body has been in a 74° room since its death, what was the time of death?

28. In 1976, the cost of a ticket to the movies was $1.75. The cost has been increasing exponentially since then and in 1998 the same ticket cost $6.50. If the cost continues to increase at this pace, how much will the ticket cost in the year 2005?

SAT PREP

1. If one "pentaminute" is the same as five minutes of time, how many pentaminutes are equivalent to four hours of time?

   a) 1200 b) 240 c) 60 d) 48 e) 20

2. If a = 12 and b = -4, what is the value of 4a - 3b?

   a) 60 b) 36 c) 16 d) 9 e) -52

3. The average (arithmetic mean) of 4 and s is equal to the average of 3, 8 and s. What does s equal?

   a) 3 b) 5.5 c) 9 d) 10 e) No such s exists.

4. In the figure at right, AB = CD. What does k equal?

   a) -6 b) -5 c) -4

   d) -3 e) -2

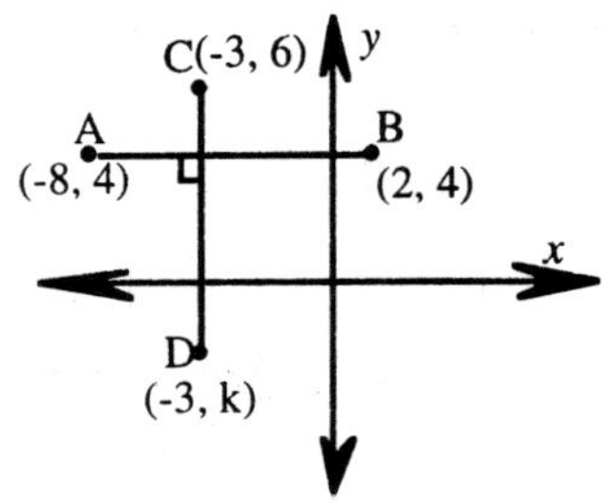

5. The initial term of a sequence is 36. Each term after that is half of the term before it <u>if</u> that term is even. If the preceding term is odd, the next term is one-half that term, plus one-half. What is the sixth term of this sequence?

   a) 1 b) 2.25 c) 3 d) 3.5 e) 4

For the next five problems, compare the items in Column A and Column B. Answer

- A if the item in Column A is greater;
- B if the item in Column B is greater;
- C if the two quantities are equal;
- D if you cannot not tell from the information given.

Note: information pertaining to both columns is given in the middle.

	Column A		Column B

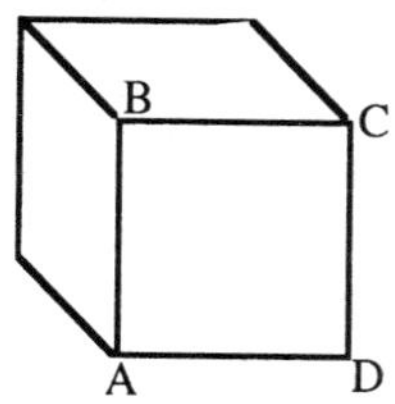

ABCD is a face of a cube.

	Column A	Column B
6.	The number of faces perpendicular to ABCD	3

m and n are integers
$m^n = 81$

	Column A	Column B
7.	m	n

The sum of the negative of p and the square of q is less than 3.

	Column A	Column B
8.	$p + 3$	q^2

$0 < g < f$

	Column A	Column B
9.	$\frac{g+1}{g}$	$\frac{f+1}{f}$

s is a positive integer
$t = 10s$

	Column A	Column B
10.	The sum of the digits of s.	The sum of the digits of t.

UNIT 7

At the County Fair
POLYNOMIALS AND GENERAL SYSTEMS

UNIT 7
POLYNOMIALS AND GENERAL SYSTEMS

This unit introduces some new ideas and information, and although these ideas might seem "complex", don't let the name fool you! Just as logarithms were invented to help mathematicians solve problems, so too were complex (or imaginary) numbers. In this unit, therefore, we will be solving equations we have seen before, but in the past have been unable to solve. In addition, we will tie this into the graphs of polynomials.

We will also look at some equations that mathematicians haven't invented a means to solve; the only way to solve these equations is by graphing or guess and check. With these equations and with the study of polynomials we will be doing, your graphing calculator needs to be close by throughout this unit!

Let's start with a quick review of some important terms so that you'll know what I'm talking about. A polynomial is an expression that involves simple operations (addition, subtraction, multiplication), and in which the variable is raised to positive integer powers. Quadratics are polynomials, but exponentials are not (since the variable is the exponent). Try this next problem to see if you can recognize a polynomial.

1. State whether or not each of the following is a polynomial. If it is not, explain why not.

a) $-7x^4 + \frac{2}{3}x^3 + x^2 - 4.1x - 6$ **b)** $8 + 3.2x^2 - \pi x^5 - 61x^{10}$

c) $9x^3 + 4x^2 - 6x^{-1} + 7^x$ **d)** $x(x^3 + 2)(x^4 - 4)$

Solution.

The expression in part (a) is a polynomial. It is okay to have coefficients $\left(\frac{2}{3}\right)$ which are not integers. Part (b) is a polynomial as well. The order of the powers of the variable is not important, nor does it matter that some powers are missing (for instance, there is no x^3). Also, don't let the πx^5 bother you. π is just a number and the coefficients can be any real number. Part (c) is not a polynomial and it fails on two counts. First, the x^{-1} is not allowed in a polynomial since all exponents must be positive integers. Second, 7^x is not allowed because the variable cannot be in the exponent. The expression in part (d) is a polynomial although it may not look like one in the form it is written. Multiply the expression and you will get (after some simplification -- try it yourself! It is good practice!) $x^8 + 2x^5 - 4x^4 - 8x$.

You also need to know what the **degree** of a polynomial is. The degree is the highest exponent of the variable. In part (a) of the previous example, the degree is 4, in part (b) the degree is 10, and in part (d) the degree is 8. Part (c) has no degree since it isn't a polynomial.

Another term you will hear a lot in this unit is **roots**. This is just another way to refer to the **x-intercepts.** In the graph at right, the parabola, g(x), has two roots: x = 3 and x = 5. The other curve, f(x), appear to have only one root: x = -6.

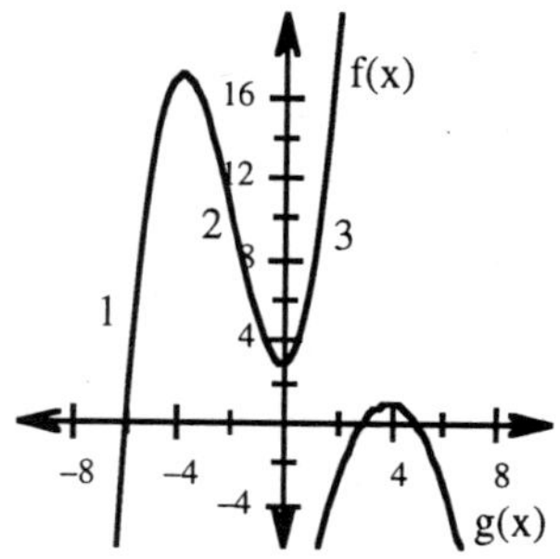

With polynomials, the degree, the graph and the number of roots are all closely related. It is this relationship we will focus on. The degree of a polynomial tells you the maximum number of times the graph of the polynomial turns, or changes direction. A fifth degree polynomial will change direction at most five times. Look at f(x) at right again. Moving from left to right, the curve goes up (1), then down (2), then back up again (3). The curve changes direction three times so it is most likely the graph of a polynomial of degree three. (I do have to put in that phrase "most likely" as there are times that you can't be sure. You saw this in CF-26. Take a look at that problem again for further explanation.) The degree also tells us the maximum number of roots a polynomial can have. Sometimes it is easy to see exactly what the roots are. For instance, if the polynomial is written in factored form, such as $y = (x - 1)(x + 6)(x - 7)$ then we can easily see what the roots of the polynomial are. This polynomial, $y = (x - 1)(x + 6)(x - 7)$, has roots x = 1, x = -6, and x = 7.

You might be realizing that the degree and roots can give us information about the graph of the polynomial. This is true. In fact, with just two other pieces of information, you can get a very good graph of a polynomial. The first additional piece of information is that of orientation. Just as $y = -x^2$ is the "upside down" version of $y = x^2$, other polynomials can have a "negative" orientation as well. For instance, the graph above of f(x) is the positive orientation of a cubic (degree 3) polynomial. The graph at right is the negative orientation, that is, it is the graph of -f(x).

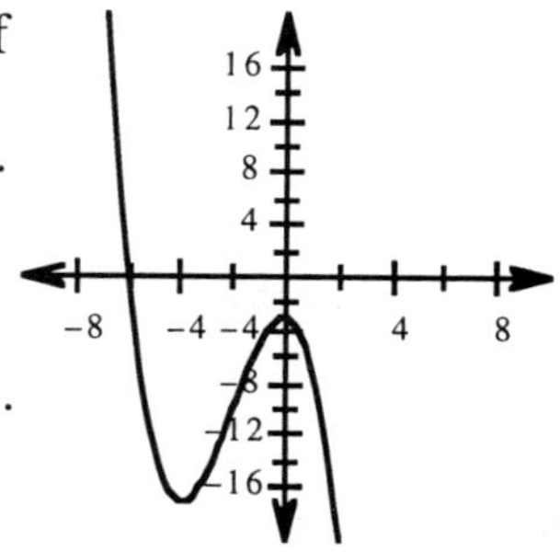

The second piece is knowing what a repeated root is. Some polynomials have double or even triple roots as you saw in CF-27. The polynomial $y = (x + 1)^2(x - 3)$ has a double root at x = -1. The graph just touches the x-axis at the double root and bounces off without passing through it.

Let's use all of these ideas in a few examples. Try these on your own first.

2. State the degree of each of the following polynomials. Then without using your graphing calculator, make a <u>quick</u> sketch of each one by using the orientation, roots and degree.

a) $f(x) = (x + 1)(x - 3)(x - 4)$ **b)** $y = (x - 6)^2(x + 1)$

c) $P(x) = x(x + 1)^2(x - 4)^2$ **d)** $f(x) = -(x + 1)^3(x - 1)^2$

Solution.
To find the degree of a polynomial when it is in factored form, you don't have to multiply the expression out completely. Remember that the degree is the highest exponent on the variable, so we don't need to know anything about the coefficients in a polynomial in order to determine its degree. Therefore, to find the degree we only need to find the exponent when all the variables are multiplied together. In part (a), therefore, the degree is 3 because $x(x)(x) = x^3$. In part (b), remember that $(x - 6)^2 = (x - 6)(x - 6)$; thus, we find the degree is also 3 because $(x^2)(x) = x^3$. The degree of P(x) in part (c) is 5 because $x(x^2)(x^2) = x^5$. Lastly, the degree of f(x) in part (d) is also 5 because $x^3(x^2) = x^5$.

To make a quick sketch of each of these functions, I will use the roots which are all easy to see since the polynomials are in factored form. Part (d) is the only graph with a negative orientation (note the negative in front of the first parentheses) but parts (b), (c) and (d) all have double roots and part (d) has a triple root as well. Here are the sketches.

a) $f(x) = (x + 1)(x - 3)(x - 4)$

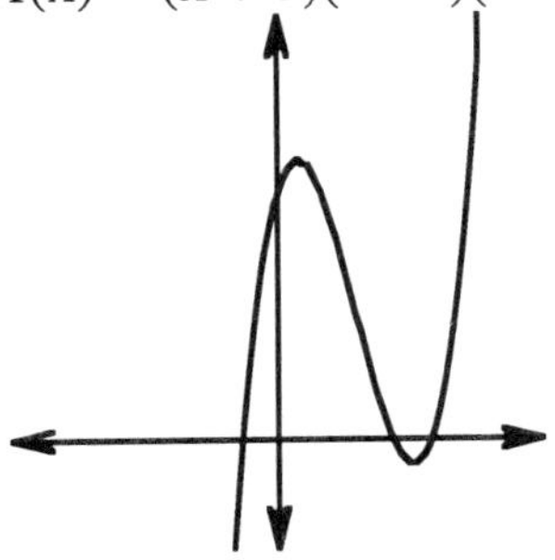

b) $y = (x - 6)^2(x + 1)$

c) $P(x) = x(x + 1)^2(x - 4)^2$

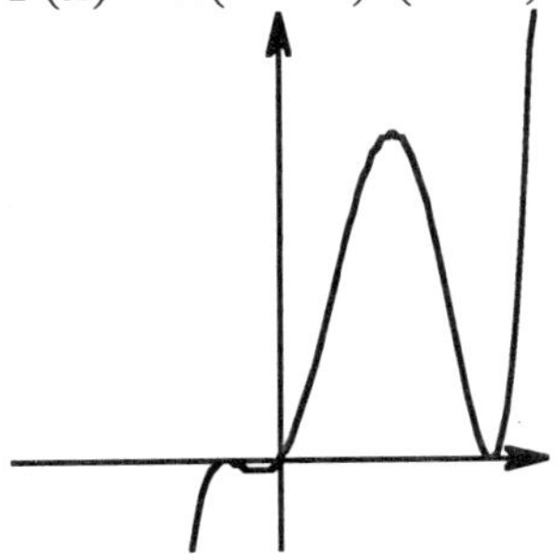

d) $f(x) = -(x + 1)^3(x - 1)^2$

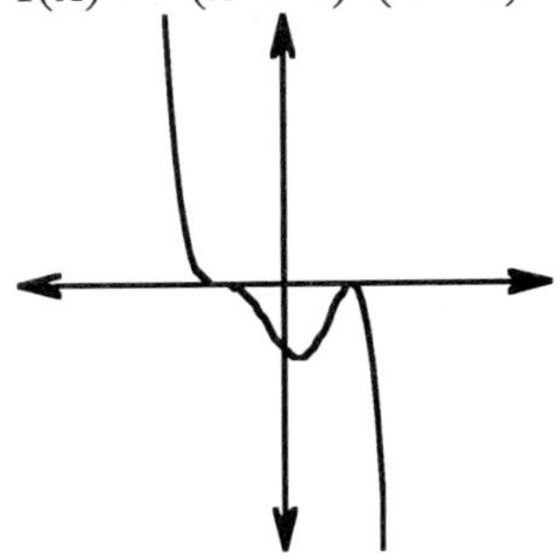

Note: these are <u>sketches</u> so don't worry if your graphs are not identical to mine. You do want to have the same general shape (number of turns), orientation and number of roots. Part (d) is particularly tricky because it has a triple root. As you found in CF-27, a triple root gives the curve a little "kink" in it, a turn similar to that of $y = x^3$ at the origin.

Knowing as much as you can about the graph of polynomials and other functions will help you solve some perplexing equations. Consider for instance the following example. Solve it on your own first.

3. Solve the following equation for x: $3^x = 4x^2 - 4$

Solution.

This may surprise you but there is no algorithm (step-by-step method) for solving this sort of equation exactly. You have to use your brain for this one. To see how we are going to solve this, let's look back at a simpler problem. When you wanted to find where the graphs of $y = 3x + 5$ and $y = 2x - 6$ crossed, you could graph them and find the point of intersection. But, you could also have found the point of intersection algebraically by substituting as $3x + 5 = 2x - 6$ and solving for x. For our new equation, we can use this same idea, but in reverse. Since we cannot algebraically solve the equation $3^x = 4x^2 - 4$, let's rewrite it as two functions to be graphed: $y = 3^x$ and $y = 4x^2 - 4$. Next use your graphing calculator to find where the curves cross. Of course, we will have to do some zooming in and some guessing and checking, but we should be able to find a solution accurate to several decimal places.

The graph of each of these functions is shown at right. Notice that they appear to cross in two places. This means our equation probably has two solutions. As is, we can see that one solution (at point A) is a little less than -1. The other at point B looks like x is a little larger than 1.5. To make these estimates more accurate we zoom in at the points of intersection, one at a time, and use the trace button to find a better approximation for x. By doing this, we can find $x \approx -1.039$ fairly quickly. At point B, $x \approx 1.53$ is a quick approximation. Try to get more accurate yourself. To five decimal places, x is approximately -1.03915 and 1.53104.

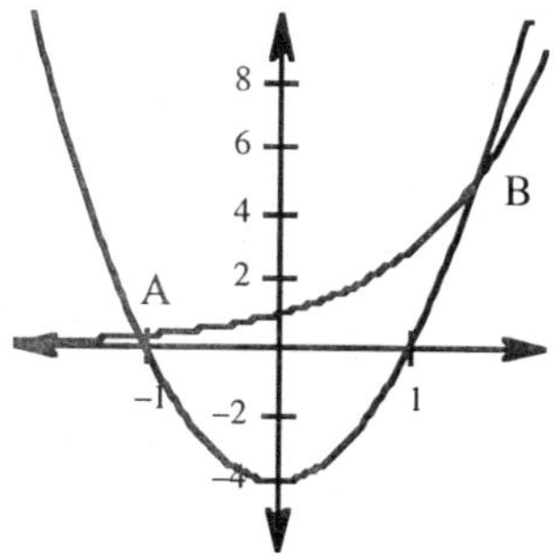

But be careful! We have missed one! Exponential functions are very steep and in fact, these curves cross again past $x = 3$. Zoom out with your calculator to verify this. By tracing and zooming, we can find this third solution to be $x \approx 3.41799$.

Before you begin to think that all equations will be this time consuming and intricate to solve, there are plenty of systems of equations that we will be able to solve using previously-learned methods. I already briefly mentioned substitution above, but also the addition and subtraction methods will be helpful. The big difference between the systems of equations we look at in this unit and the ones we have already seen is that these systems will not be just the intersections of lines. In this unit we will let lines intersect parabolas, lines intersect circles or even circles intersect parabolas. The solution methods will be similar to what we have done earlier in the course, but we will have more than one solution to many of them. Let's try this below.

4. Solve the following systems of equations. Use the graphs of each equation to verify and explain your solutions. Be sure you have all possible solutions!

a) $y = x + 3$
$y = (x + 4)^2 - 7$

b) $x^2 + y^2 = 25$
$y = x^2 - 5$

c) $(x + 2)^2 + y^2 = 36$
$y = -(x + 2)^2 + 16$

d) $y = (x - 3)^2 + 4$
$y = -x + 2$

Solution.
Before you start to solve these problems, it helps to think about what these are graphs of so you know what to expect. For instance, in part (a), the first equation is a line and the second equation is a parabola. A line and a parabola can intersect in at most two points, but they can intersect in only one point or not intersect at all. Now let's solve each of these to see what happens.

In part (a), I will use substitution. (This is often done when one, or both, equations are written as "y =".)

$$\left.\begin{aligned} y &= x + 3 \\ y &= (x + 4)^2 - 7 \end{aligned}\right\} \quad \rightarrow \quad x + 3 = (x + 4)^2 - 7$$

Now we are left with a quadratic equation to solve.

$$x + 3 = (x + 4)^2 - 7$$

$$
\begin{aligned}
x + 3 &= x^2 + 8x + 16 - 7 \\
x + 3 &= x^2 + 8x + 9 \\
0 &= x^2 + 7x + 6 \\
0 &= (x + 6)(x + 1) \\
x &= -6, -1
\end{aligned}
$$

We are not done yet. We have only found the x values. We still need to find the corresponding y-values. For this we can use either equation. I will use the simpler one: $y = x + 3$.

$$
\begin{aligned}
y &= -6 + 3 = -3, \text{ and} \\
y &= -1 + 3 = 2
\end{aligned}
$$

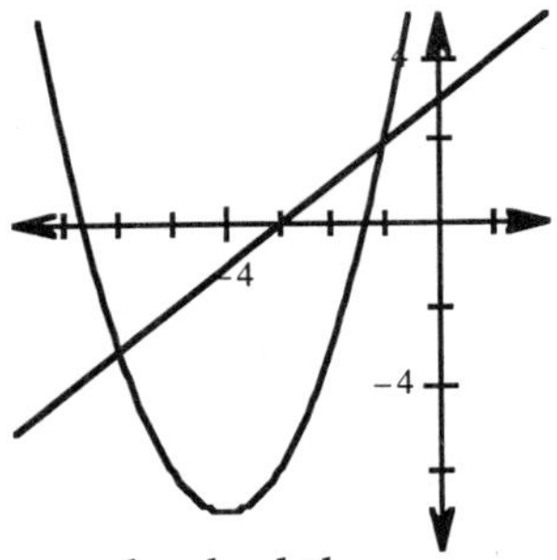

Our solution appears to be (-6, -3) and (-1, 2). Make sure you pair up the correct coordinates! Lastly, we need to examine the graphs to see if this solution seems reasonable. The graphs are shown at right and as you can see, they certainly intersect in two points, and only two points, so we don't have to look for other solutions. Also, the points we found seem reasonable based on the graphs. Note: you can (and should) check your solutions algebraically by substituting the x and y values back into both equations to see that they do check. Just as before, this is something you should always do, especially on tests! Although I have checked these points algebraically as well, I am not showing my check here because it is too lengthy.

Let's move on to part (b). Again I will use substitution. The second equation has already been solved for y so the natural thing to do might be to replace the y in the first equation with $x^2 - 5$. But doing that will mean you will need to solve this equation:

$$x^2 + (x^2 - 5)^2 = 25$$

When multiplied out, this equation is a fourth degree polynomial. That would be a bit more difficult to solve then what we are used to. Instead, I will use the second equation and solve for x^2:

$$y = x^2 - 5 \quad \rightarrow \quad y + 5 = x^2$$

Now replace x^2 in the first equation with $y + 5$.

$$y + 5 + y^2 = 25$$

This is just a quadratic equation. Solve it by the Quadratic Formula or see if it factors.

$$
\begin{aligned}
y^2 + y - 20 &= 0 \\
(y + 5)(y - 4) &= 0 \\
y &= -5, 4
\end{aligned}
$$

Now that we have y, we must find the corresponding x values. I'll use the second equation to do this.

$$
\begin{aligned}
-5 &= x^2 - 5 \\
0 &= x^2 \\
x &= 0
\end{aligned}
\qquad \text{and} \qquad
\begin{aligned}
4 &= x^2 - 5 \\
9 &= x^2 \\
x &= \pm 3
\end{aligned}
$$

The solutions appear to be (3, 4), (-3, 4), and (0, -5). The graph of these is at right and as we can see this parabola and circle really do only intersect in three points. This is because the parabola is tangent to the circle at its vertex.

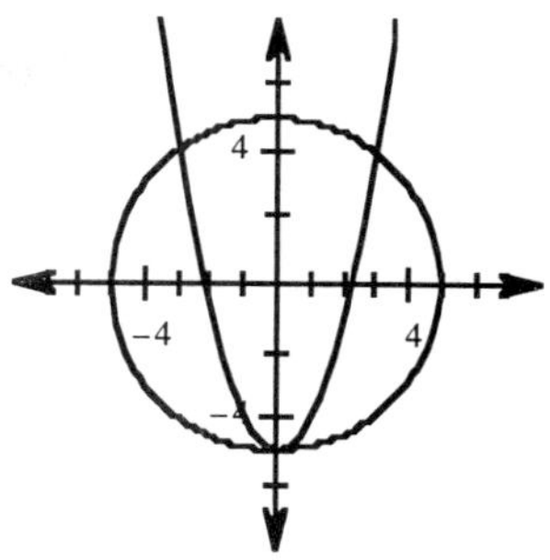

In part (c) we have another parabola and circle. Imagine a parabola and circle intersecting. They won't always produce three intersection points as in the last example. They could intersect in one point, two points, three points (as we just saw), four points, or even <u>no</u> points. To begin, I will solve the second equation for $(x + 2)^2$:

$$
\begin{aligned}
y &= -(x + 2)^2 + 16 \\
y - 16 &= -(x + 2)^2 \\
-y + 16 &= (x + 2)^2
\end{aligned}
$$

Substitute this expression for $(x + 2)^2$ in the first equation and solve for y:

$$
\begin{aligned}
-y + 16 + y^2 &= 36 \\
y^2 - y - 20 &= 0 \\
(y - 5)(y + 4) &= 0 \\
y &= 5, -4
\end{aligned}
$$

Now find x (are you starting to see a trend?):

$$
\begin{aligned}
5 &= -(x + 2)^2 + 16 \\
-11 &= -(x + 2)^2 \\
11 &= (x + 2)^2 \\
\pm\sqrt{11} &= x + 2 \\
-2 \pm \sqrt{11} &= x, \text{ and}
\end{aligned}
$$

$$
\begin{aligned}
-4 &= -(x + 2)^2 + 16 \\
-20 &= -(x + 2)^2 \\
20 &= (x + 2)^2 \\
\pm\sqrt{20} &= x + 2 \\
-2 \pm 2\sqrt{5} &= x
\end{aligned}
$$

The solutions are (being careful to pair up the correct x and y values of course) $(-2 + 2\sqrt{5}, -4)$, $(-2 - 2\sqrt{5}, -4)$, $(-2 + \sqrt{11}, 5)$, and $(-2 - \sqrt{11}, 5)$. The graph confirms these points. Note: In cases such as these, it is probably best to approximate the answers (i.e., rewrite expressions such as $-2 + 2\sqrt{5}$ in its decimal approximation: 2.4721). This will let you check the position of the points more easily than writing them in radical form.

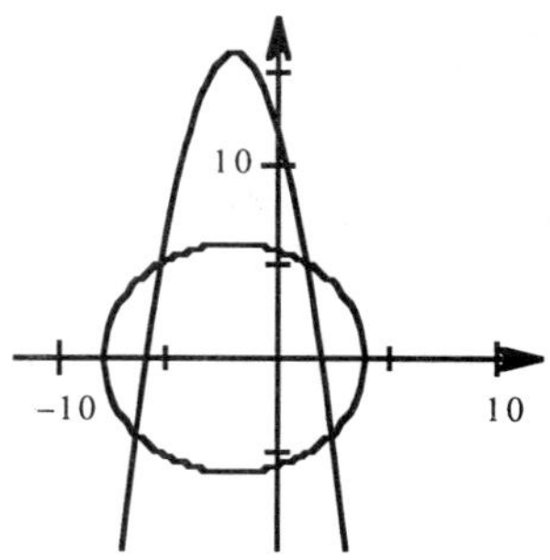

Notice that the substitutions in these problems are more sophisticated than those in simpler problems like linear systems. Spend some time planning your solution process and many of these problems will be easier to solve.

Next we solve part (d). Since both equations are written as "y =" I will use a straightforward substitution and solve for x:

$$\left.\begin{array}{r} y = -x + 2 \\ y = (x - 3)^2 + 4 \end{array}\right\} \quad \rightarrow \quad -x + 2 = (x - 3)^2 + 4$$

$$\begin{aligned} -x + 2 &= (x - 3)^2 + 4 \\ -x + 2 &= x^2 - 6x + 9 + 4 \\ 0 &= x^2 - 5x + 11 \end{aligned}$$

We will use the Quadratic Formula since this equation does not factor.

$$x = \frac{5 \pm \sqrt{(-5)^2 - 4(1)(11)}}{2(1)}$$

$$x = \frac{5 \pm \sqrt{25 - 44}}{2}$$

$$x = \frac{5 \pm \sqrt{-19}}{2}$$

We have a negative under the radical sign I've checked my work and I haven't made a mistake, so what does $\sqrt{-19}$ tell us? Let's look at the graph for some help.

The graph of the parabola and line are shown at right. These two curves do not intersect at all, so it shouldn't be surprising that we couldn't find any real solutions. Therefore, the solution to part (d) is <u>no solution</u>.

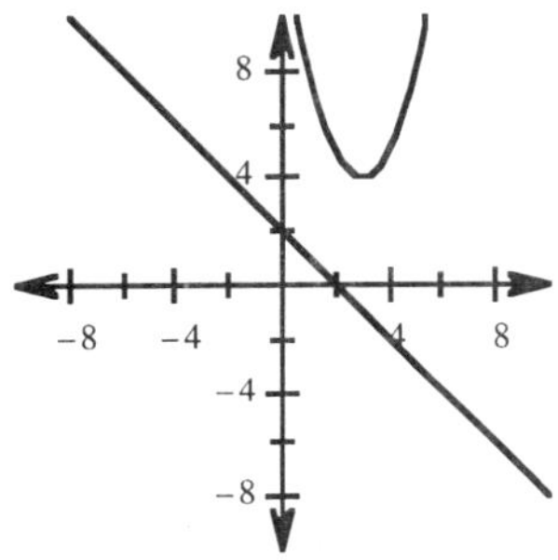

One thing I should point out: my solutions are oftentimes very long. Don't forget that I'm trying to explain what I am doing so that other people can learn from it. I also try to put in helpful tips and thoughts I have while doing the problem. Your solutions, even when you are very careful and show all your work, will usually not be as long as mine.

In the last problem, we saw that curves can intersect in various ways and can, in fact, not intersect at all as in part (d). It is the last case that we will look at next.

When we have two curves that do not intersect, solving them poses a problem. We have seen that mathematicians don't like equations that can't be solved and will invent new means in order to solve them (e.g., creating logarithms in order to solve equations such as $3^x = 4$). Guess what? They did it for these systems as well. The problem lies within the square root. Until now, we have not been able to take the square root of a negative number, but the introduction of a new number will solve the problem. Mathematicians defined $\sqrt{-1}$ to equal **i**, an **imaginary** number. This lets us find the square root of any negative number. For instance, \r(-19) = \r(-1) $\sqrt{19} = \mathbf{i}\sqrt{19}$, and $\sqrt{-25} = \sqrt{-1}\,\sqrt{25} = \mathbf{i} \cdot 5 = 5\mathbf{i}$. Sometimes when we solve these equations, we don't just get an imaginary number, we get a **complex** number, the sum of a real number and an imaginary number. Numbers such as $6 + 5\mathbf{i}$ and $-3 - 2\mathbf{i}$ are complex numbers and you will need to be able to perform simple operations on these numbers. Try the next problems to see how you do.

5. Simplify the following.

a) $3 + \sqrt{-16}$ **b)** $\dfrac{6 + \sqrt{-9}}{3}$ **c)** $(3 + 4i) + (-2 - 6i)$

d) $(4i)(-5i)$ **e)** $(-7 + 2i)(5 + i)$ **f)** $(8 - 3i)(8 + 3i)$

Solution.
When simplifying complex numbers, there are a few key things to remember. First, remember that you can only add or subtract like terms. This means real parts add to real parts and imaginary parts add to imaginary parts. The second thing to remember is that $\mathbf{i}^2 = -1$. With this in mind let's start simplifying.

In part (a) all we need to do is simplify the radical.

$$\begin{aligned} 3 + \sqrt{-16} &= 3 + \sqrt{16}\sqrt{-1} \\ &= 3 + 4\mathbf{i} \end{aligned}$$

Part (b) gives us two tasks: simplify the radical and then checking to see if the fraction can be simplified.

$$\begin{aligned} \frac{6 + \sqrt{-9}}{3} &= \frac{6 + \sqrt{9}\sqrt{-1}}{3} \\ &= \frac{6 + 3\mathbf{i}}{3} \\ &= \frac{6}{3} + \frac{3\mathbf{i}}{3} \\ &= 2 + \mathbf{i} \end{aligned}$$

When adding two complex numbers as in part (c), just combine the like terms.

$$(3 + 4i) + (-2 - 6i) = 3 - 2 + 4\mathbf{i} - 6\mathbf{i} = 1 - 2\mathbf{i}$$

There is less to worry about when multiplying complex numbers. Just remember what you can and cannot do with variables. We cannot simplify x + y, it just stays x + y, but we can multiply them: (x)(y) = xy. This is true of complex numbers as well.

$$(4\mathbf{i})(-5\mathbf{i}) = -20\mathbf{i}^2 = -20(-1) = 20$$

Remember! $\mathbf{i}^2 = -1$! As you can see, it is possible to end up with a completely <u>real</u> answer. Parts (e) and (f) are similar and are the type you will see most often. Multiplying two complex numbers as we do here is the same as multiplying two binomials. So, whether you use rectangles to multiply binomials or use FOIL, your method will work on multiplying complex numbers, too.

If you find you make mistakes when multiplying complex numbers, be sure to show your work and <u>don't</u> try to do the multiplication in your head! For part (e) I have used a rectangle shown at right. Combining like terms gives

	-7	2i
5	-35	10i
i	-7i	$2i^2$

$$-35 + 3\mathbf{i} + 2\mathbf{i}^2 = -35 + 3\mathbf{i} - 2 = 37 + 3\mathbf{i}$$

I will multiply the expression in part (f) without rectangles. You may wish to still use them.

$$\begin{aligned}(8 - 3\mathbf{i})(8 + 3\mathbf{i}) &= 64 + 24\mathbf{i} - 24\mathbf{i} - 9\mathbf{i}^2 \\ &= 64 - 9\mathbf{i}^2 \\ &= 64 + 9 \\ &= 73\end{aligned}$$

You will see a real number result happen fairly often. This happens when the two complex numbers are **conjugates** of each other. Look at the two complex numbers: they are the same except for the sign of the imaginary part. Watch for these pairs. recognizing them and knowing the pattern that simplifies them usually make your work easier.

So how do complex numbers and polynomials tie together? Through the roots. Remember I said that the term "roots" is just another way to refer to the x-intercepts of a polynomial. That is actually only part of it. Take the function $y = x^2 + 5$, a simple equation with a simple graph. As you can see, this parabola does not have any x-intercepts because it does not cross the x-axis. But it <u>does</u> have roots! Let's try setting the quadratic equal to zero and solving, as we do to find the x-intercepts, and see what happens.

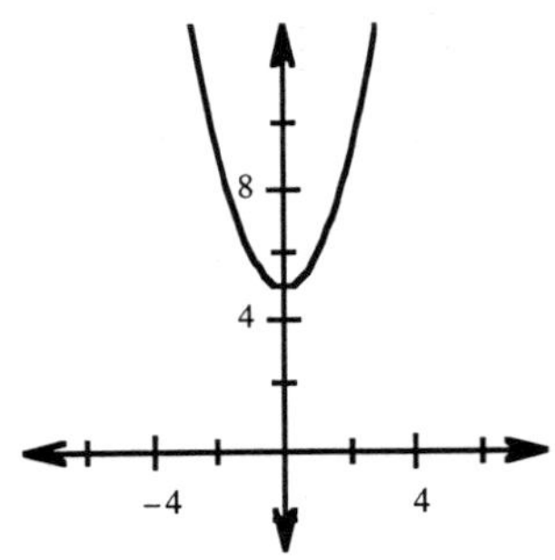

$$\begin{aligned}x^2 + 5 &= 0 \\ x^2 &= -5 \\ x &= \pm\sqrt{-5} \\ x &= \pm 5\mathbf{i}\end{aligned}$$

In fact, this equation does have roots, they are just complex! This second degree equation has two complex roots: $5\mathbf{i}$ and $-5\mathbf{i}$.

Earlier I showed you the graph at right and called it f(x). We deduced by the number of times the graph changes direction that this was probably a degree 3 polynomial. In fact it is. How many roots does this polynomial have? It has only one x-intercept at $x = -6$. Let's look at the equation.

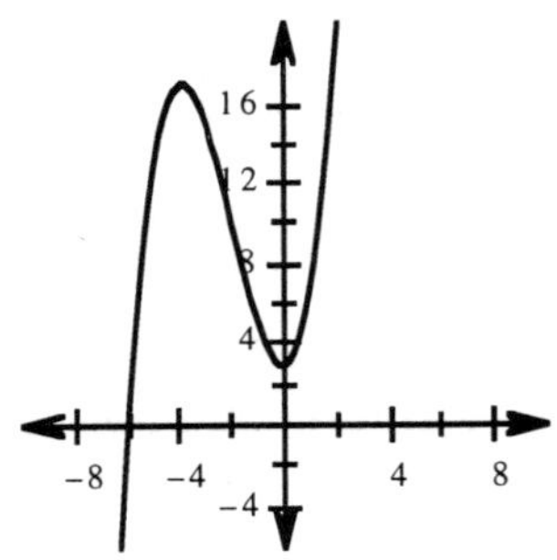

$$f(x) = 0.5(x + 6)(x^2 + 1)$$

The equation confirms that this is degree 3. What if we tried to solve for the x-intercepts? Let's set this equal to zero and solve for x.

$$0 = 0.5(x + 6)(x^2 + 1)$$

The result is a third degree polynomial, but it is in factored form so we can quickly and easily use the Zero Product Property to solve for the roots.

$$\begin{aligned}x + 6 &= 0 \\ x &= -6\end{aligned} \qquad \text{or} \qquad \begin{aligned}x^2 + 1 &= 0 \\ x^2 &= -1 \\ x &= \pm\sqrt{-1}\end{aligned}$$

We end up with one real root: $x = -6$ and two imaginary roots: $x = \pm\,\mathbf{i}$. This is actually confirmed by the graph and the degree. The degree of a polynomial also tells us the number of roots the polynomial must have. Some of those roots though may not be real as we just saw.

Below are several graphs of polynomial functions. Test your knowledge on what we have discussed so far.

6. For each graph below, state the degree of the polynomial function and the number and type of roots each has. Assume the graphs are complete.

a) b) c)

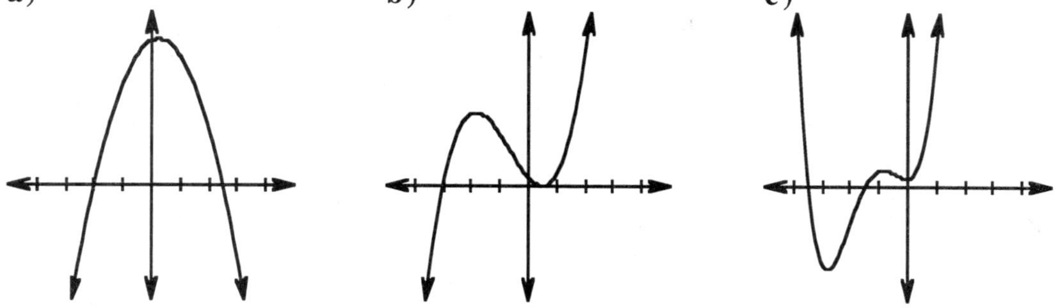

Solution.
The first polynomial appears to be a parabola so it is a degree two polynomial with two real roots (x-intercepts). The curve in part (b) has three turns so it is probably a degree three. Its roots are all real, but there seems to be only two of them. That is because the second one (at approximately x = 0.5) is a double root. (Review CF-27 if you need to). The third graph has four turns so we would expect it to be a degree four. We can see two real roots (at $x \approx -3.5$ and $x \approx -1.5$) so the other two must be complex to give a total of four roots.

You might have noticed already but complex roots always come in pairs. In fact, they always come with their complex conjugate pair. This means if you know that one root of a polynomial is 3 + 7**i**, then 3 - 7**i** must be another root. This can help you find roots and the equations of polynomial functions. Let's see how in a quick example.

7. Three of a polynomial function's roots are 2, 1 - 4i and 3i. What is the minimum degree of this function and one possible equation of this function?

Solution.
One of the roots is real, x = 2. We will leave that one alone for now. Real roots don't come in pairs. The other two are complex and complex roots must come in pairs. Since 1 - 4**i** is a root, its conjugate, 1 + 4**i** must also be a root. Also, since 3**i** is a root, its conjugate, -3**i** must also be a root. Now we have five roots: 2, 1 - 4**i**, 1 + 4**i**, 3**i**, and -3**i**. This polynomial function must be at least degree five.

How do we find a possible equation? Let's remember a simpler, similar problem. If we knew a parabola had x-intercepts at x = 3 and x = -2, then we would know the factored form of the parabola would be (x - 3)(x + 2). (Think about solving the factored form with the Zero Product Property.) Similarly, if our fifth degree polynomial has 2 as a root, then x - 2 is a factor. If 1 - 4**i** is a root, then x - (1 - 4**i**) is a factor. If 1 + 4**i** is a root, then x - (1 + 4**i**) is a factor. Getting the idea? We can write down a factored form of a possible equation just by using the roots.

$$f(x) = (x - 2)(x - (1 - 4\mathbf{i}))(x - (1 + 4\mathbf{i}))(x - 3\mathbf{i})(x + 3\mathbf{i})$$

Before you multiply this out, take this advice: don't multiply it out unless your teacher specifically tells you to multiply, simplify, or write with only real number constants. Second piece of advice: if you must multiply it out, multiply conjugate pairs together first. You might remember that when we multiply conjugate pairs, they simplified the expressions greatly. If you'd like the practice, try multiplying the above expression out. Watch your negative signs. You should get
$f(x) = (x - 2)(x^2 - 2x + 17)(x^2 + 9) = x^5 - 4x^4 + 30x^3 - 70x^2 + 189x - 306$ when it is simplified.

Next we want to come up with the exact equations to fit the polynomial curves. We will do this by using the roots and one other point.

8. For each graph below, write a specific equation.

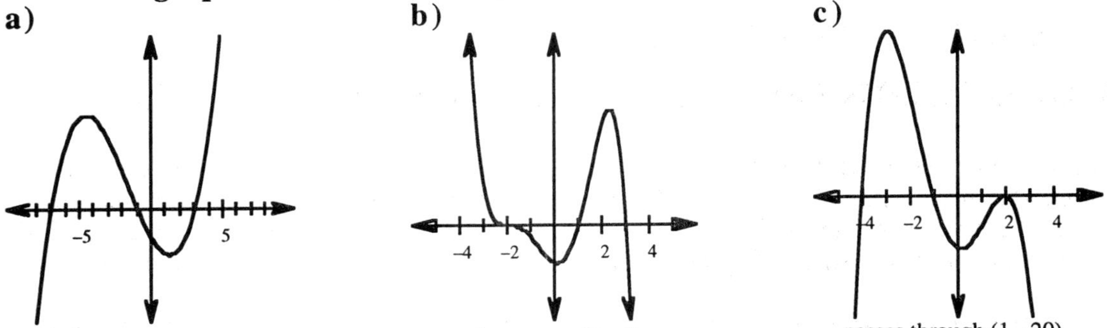

Solution.

For each of these functions we will use the roots to write the factors of the polynomial so that we will get an equation which looks like $y = a(x - r_1)(x - r_2)(x - r_3) \ldots$ where r_1, r_2, r_3, ... are the roots. Then we will use the extra point given to find the value of a.

The roots in part (a) are x = -7, x = -1, and x = 3. We can use the factors to write a general equation y = a(x + 7)(x + 1)(x - 3). To determine the value of a, let's use the y-intercept (0, -7). Substitute x = 0 and y = -7 into this equation and solve for a.

$$\begin{aligned} -7 &= a(0 + 7)(0 + 1)(0 - 3) \\ -7 &= a(7)(1)(-3) \\ -7 &= -21a \\ a &= \frac{1}{3} \end{aligned}$$

Therefore the specific equation for the graph in part (a) is $y = \frac{1}{3}(x + 7)(x + 1)(x - 3)$.
Note: I was not asked to simplify or multiply this out so I'm not going to.

Now do the same thing for part (b). The roots are x = -2, x = 1, and x = 3. The first root, because of its "s" turn is a triple root. The general equation is
$y = -a\,(x + 2)^3(x - 1)(x - 3)$. To find the value of a we will use the point (0, -6).

$$-6 = a(0 + 2)^3(0 - 1)(0 - 3)$$
$$-6 = a(8)(-1)(-3)$$
$$-6 = 24a$$
$$a = -\frac{1}{4}$$

Therefore the specific equation is $y = -\frac{1}{4}(x + 2)^3(x + 1)(x - 3)$. Again I won't simplify!

Lastly, part (c) has roots $x = -4$, $x = -1$, and $x = 2$ which is a double root. The general equation for this curve then is $y = a(x + 4)(x + 1)(x - 2)^2$. Here we are not given the y-intercept but another point (1, -20). Just as with the others, I'll substitute these values in for x and y in order to solve for a.

$$-20 = a(1 + 4)(1 + 1)(1 - 2)^2$$
$$-20 = a(5)(2)(1)$$
$$-20 = 10a$$
$$a = -2$$

For part (c) then the specific equation is $y = -2(x + 4)(x + 1)(x - 2)^2$.

If you find you must simplify these polynomial equations, there is a way to check your work. On the same set of axes, use your graphing calculator to graph the factored form and the simplified form. The curves should lie right on top of each other. In fact, it should look like there is only one curve there. If you see two different curves, go back and check your work.

Many of the topics in this unit are the beginnings of serious mathematics. If you can see how the ideas are related, not only will you understand this unit better, you will understand other units, other math courses, and other math-related courses better. As with the other units, I have not covered every type of problem discussed in the unit. But what is here will give you assistance with comprehending everything else done in this unit.

This is a good time to reflect on the course so far, as asked in CF-147. Take the time to not only reflect on what you have learned, but what you are planning to do with what you have learned. Undoubtedly your teachers have told you time and time again about all the jobs that require some mathematics. What do you think you will do with your knowledge?

UNIT 7
MORE TO TRY

State whether or not each of the following is a polynomial. If it is, give the degree. If it isn't, explain why not.

1. $\frac{1}{8}x^7 + 4.23x^6 - x^5 + 17x^4 - \pi x^3 + \sqrt{2}\,x^2 - .01x - 12.4$

2. $45x^3 - \frac{3}{4}x^2 - .03x + \frac{5}{x} + 15$

3. $x(x + 2)\left(6 + \frac{1}{x}\right)$

Sketch a graph of each of the following.

4. $y = (x + 5)(x - 1)^2(x - 7)$

5. $y = -(x + 3)(x^2 + 2)(x + 5)^2$

6. $f(x) = -x(x + 8)(x + 1)$

7. $y = x(x + 4)(x^2 - 1)(x - 4)$

Solve the following equations accurate to three decimal places. Briefly justify your answer.

8. $5^x = -3x + 6$

9. $3(1.3^x) = x^2 - 2$

10. $\log x = -(x - 4)^2 + 3$

Simplify the following.

11. $(6 + 4\mathbf{i}) - (2 - \mathbf{i})$

12. $8\mathbf{i} - \sqrt{-16}$

13. $(-3)(4\mathbf{i})(7\mathbf{i})$

14. $(5 - 7\mathbf{i})(-2 + 3\mathbf{i})$

15. $(3 + 2\mathbf{i})(3 - 2\mathbf{i})$

16. $(\sqrt{3} - 5\mathbf{i})(\sqrt{3} + 5\mathbf{i})$

Solve the following systems. Use the graphs to verify your answers.

17. $x^2 - 6y = 34$
 $x^2 + y^2 = 25$

18. $y = -2x^2 + 8x + 27$
 $2x + y = 15$

19. $x^2 + y^2 = 64$
 $x^2 + 10y = 100$

20. $x^2 + y^2 = 36$
 $x + y = 4$

Below are the <u>complete</u> graphs of some polynomial functions. Based on the shape and location of the graph, describe all the roots of the polynomial function. Be sure to include information like whether roots are double or triple, real or complex, etc.

21.

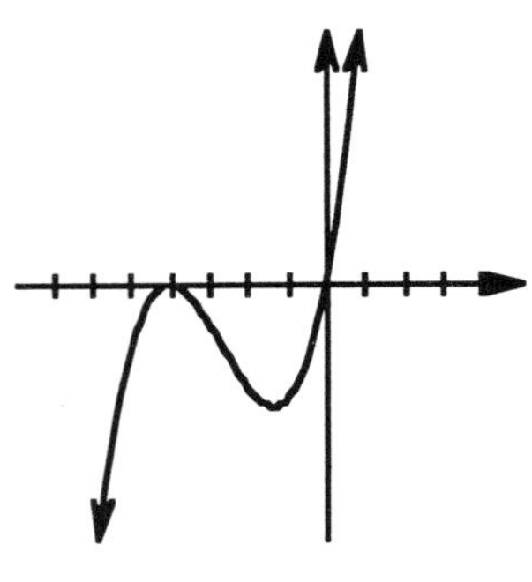

22.

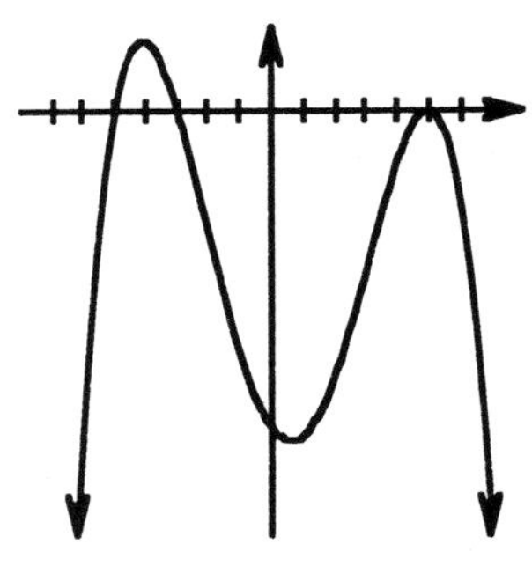

23.

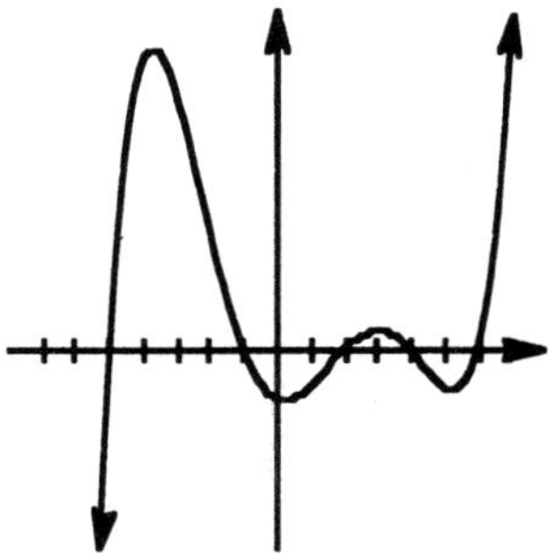

24.

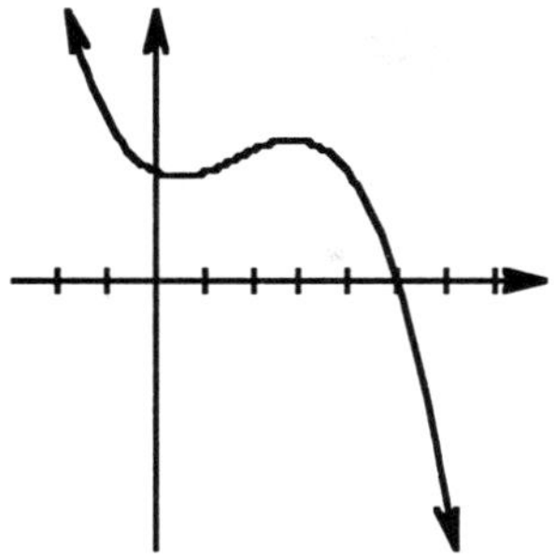

25.

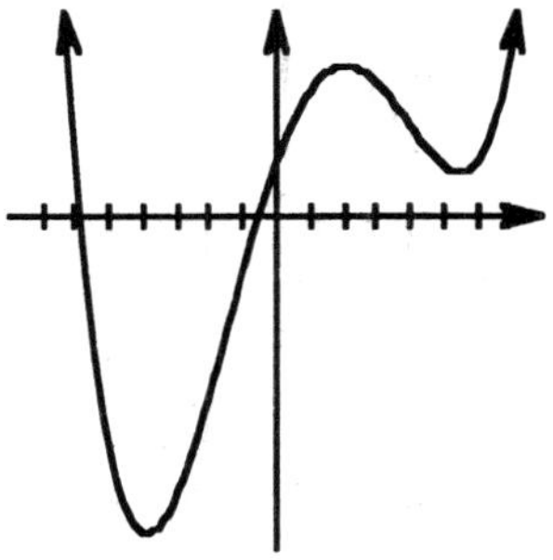

26.

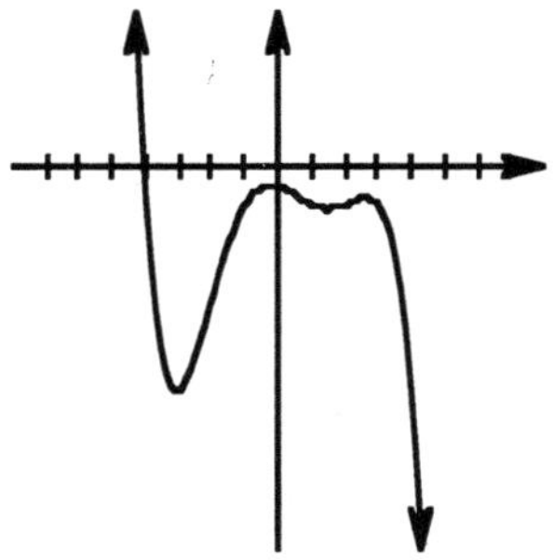

Using all the information supplied to you, write the specific equation for each polynomial function.

27. y -intercept: (0, 12)

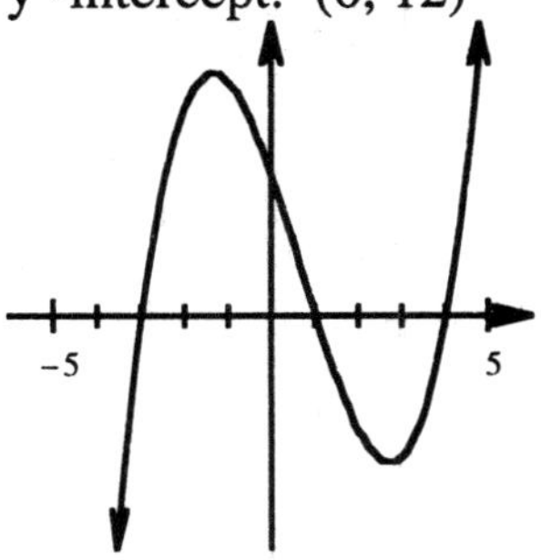

28. y-intercept: (0, -15)

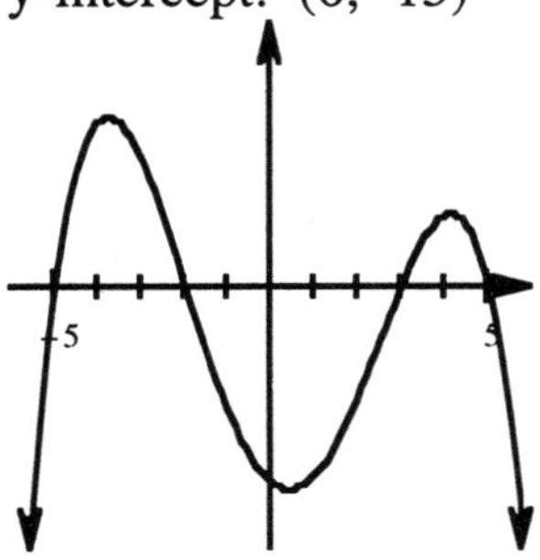

29. y-intercept: (0, 3)

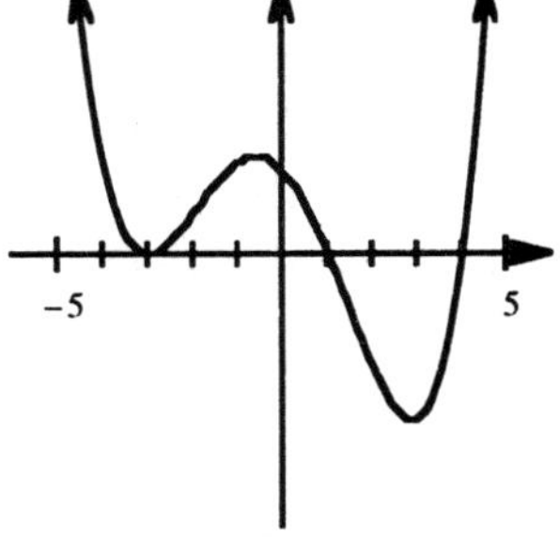

30. Passing through the points (0, 5), and with roots x = 5, x = -2, and x = 3**i**.

SAT PREP

1. What number decreased by five is four times the number?

   a) $\frac{5}{4}$ b) 1 c) 0 d) -1 e) $\frac{-5}{3}$

2. The distance from Public School 1 (PS1) to Public School 2 (PS2) is six miles. The distance from PS2 to PS3 is eight miles. Which of the following could NOT be the distance from PS1 to PS3?

   a) 15 b) 14 c) 12 d) 3 e) 2

3. If $x \neq 0$ then $\dfrac{1}{\frac{8}{4x}} =$

   a) 4x b) 2x c) $\frac{x}{2}$ d) $\frac{2}{x}$ e) $\frac{x}{4}$

4. In the figure at right, the circles are concentric with OA = 2 and AB = 5. What is the ratio of the circumference of the smaller circle to the circumference of the larger circle?

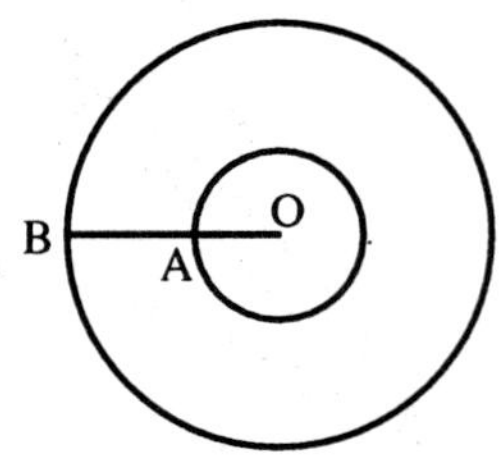

   a) $\frac{2}{5}$ b) $\frac{2}{7}$ c) $\frac{4}{25}$

   d) $\frac{4}{49}$ e) $\frac{2}{3}$

5. When grocery store items do not sell, the manager reduces the price by 10% each month. If a particular item sells for d dollars in January and does not sell till May, at what price does it sell?

   a) 0.6d b) 0 .4d c) 0.9999d d) 0.6561d e) 0.0001d

6. If .37x + 3.7, what does $\frac{1}{x}$ equal?

7. The sum of two number is four and their difference is three. What is their product?

8. If $8^7 = 2^x$, what does x equal?

9. If the perimeter of a rectangle is eight times the width of the rectangle, then the length of the rectangle is how many times the width?

10. There are 140 red marbles and 60 green marbles in a bag containing 200 marbles. If we are only to add green marbles, how many should we add so that the probability of randomly drawing a green marble is $\frac{3}{4}$?

UNIT 8

The Circle of Terror

CIRCULAR FUNCTIONS

UNIT 8
TRIGONOMETRIC FUNCTIONS

In your years of studying math, you have seen some trigonometry before now. Most likely it was in Math 2--Geometry--where you first came across sine, cosine and tangent. You probably were pretty good at solving trig problems, usually finding the lengths of sides of triangles using these functions. But this unit is different. Here we will look at sine and cosine as functions which we can graph and use to model data.

In studying trigonometry as functions, we develop some new ideas and new terms. (This will be an important time to keep your tool kit handy and up to date.) Our familiar terms will still be relevant, terms like domain, range, intercepts, etc., and many of the techniques we learned about shifting and stretching graphs will apply as well. The graphs of trigonometric functions might look different, but we can still apply what we have already learned about functions.

The Circle of Terror problem which begins the unit is important. Once you have done the problem and several problems after it, you won't have to go back and graph S(x) point by point. Also, the Circle of Terror problem is a good introduction to the unit circle and how sine (and cosine) relate to the unit circle. As the Circle of Terror problem describes, we imagine riding the circumference of a circle (the **unit circle** is a circle with radius 1 unit). Steve, the rider, sees markers every kilometer of distance he had traveled. In addition, a board displays his always changing altitude. The relationship between the distance-traveled and the altitude is a function we call **S(x)**. Let me work through one position to find the altitude.

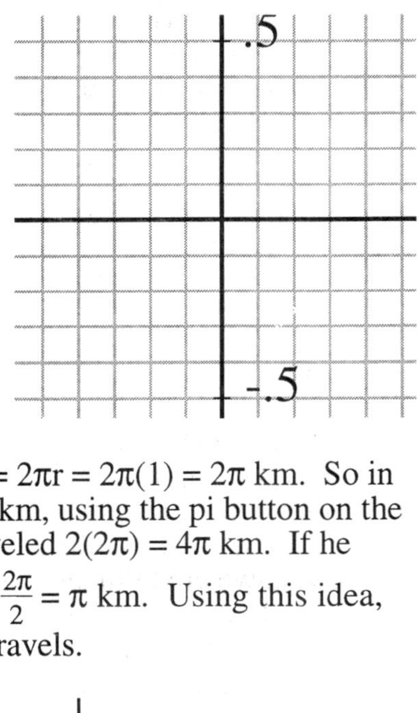

Let's look at the 2 km marker. Since the x-axis represents the ground, we can estimate the altitude of the car at this marker by estimating the y-coordinate of this marker: approximately 0.92 km. (Remember: the radius of the circle is one kilometer.) Can we find other altitudes? Sure. We can estimate as many as we'd like by using the graph.

First, let's look at the Circle of Terror again. One complete revolution is a little more than six kilometers long. How long is it exactly? What I'm really asking is "What is the circle's circumference?" Since the radius is one kilometer, the circumference (i.e. the distance traveled in one revolution) is $C = 2\pi r = 2\pi(1) = 2\pi$ km. So in one revolution, the rider has traveled 2π km, approximately 6.28 km, using the pi button on the calculator. If the rider makes another revolution, then he has traveled $2(2\pi) = 4\pi$ km. If he only makes one-half a revolution, then the rider has only traveled $\frac{2\pi}{2} = \pi$ km. Using this idea, you should be able to figure out several other distances the rider travels.

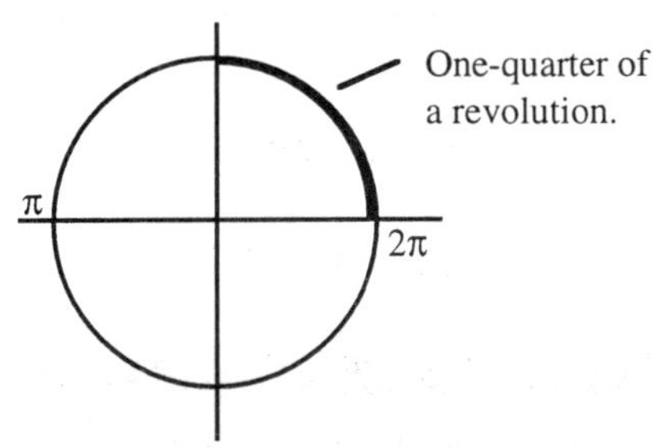

One quarter of a revolution is $\frac{1}{4}(2\pi) = \frac{\pi}{2}$. (You can also think of this as "half way to π.") Three-fourths of a revolution is $\frac{3}{4}(2\pi) = \frac{3\pi}{2}$. What if the rider only traveled one-eighth of a revolution? What about five-sixths? Let's look at these in a problem and figure out the altitude at these points. Remember: Try it first for yourself!

1. On a unit circle representing the Circle of Terror ride,

a) locate the position of the rider after one-eighth of a revolution.

b) locate the position of the rider after five-sixth of a revolution.

c) Calculate the altitudes at each point.

Solution.
We can easily cut the circle into fourths; we just did that above. To find an eighth of a revolution, we cut the first quarter in half, bisecting the 90° angle. I'll call this point A. Five-sixths of a revolution is almost a whole revolution. Dividing the circle into sixths we can travel over five of them to find our position a the point labeled B.

If the circle is accurately drawn on graph paper, we could estimate the altitude by estimating the y-coordinate for each point A and B. But, that would still only be an estimation. We need a more precise method.

Let's look at point A again. We bisected the right angle to find point A. Therefore the central angle measures 45° ($\frac{1}{2} \cdot 90°$). What we have, then, is a right triangle with hypotenuse length one and an acute angle of 45°. (Do you see why it is a right triangle? We want to find the altitude. That forms a right triangle. Do you know why the hypotenuse is one? That is the radius of the circle.) How do we find the length y? That is OLD trigonometry, stuff from Math 2. Either remember that this is a special right triangle (45°-45°-90°) or use trig functions.

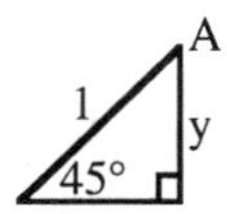

$$\sin 45° = \frac{y}{1}$$
$$y = \sin 45°$$
$$y \approx .707$$

or

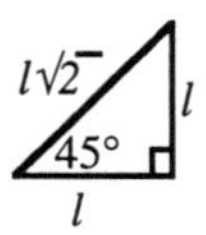

The side length pattern of an isosceles right triangle is shown at left. In this case, $l\sqrt{2} = 1$ so $l = \frac{1}{\sqrt{2}}$. If we rationalize the denominator,

$$l = \frac{1}{\sqrt{2}} \cdot \frac{\sqrt{2}}{\sqrt{2}} = \frac{\sqrt{2}}{2}$$

What about at point B? Five-sixths of a revolution means there is only one-sixth left. Can you figure out the measure of angle 1? If we know that measurement, then we are on our way to calculating the altitude. (Remember: the altitude is what we are ultimately looking for.) a complete revolution of a circle is 360°. Five-sixths of 360° is 300° (verify this yourself!), leaving 60° (one-sixth). Therefore, $m\angle 1 = 60°$. Now we can find y with special triangles (this one is a 30° - 60° - 90° triangle) or with trig functions. Using the pattern of a 30° - 60° - 90° triangle, $y = \frac{1}{2}$. Therefore the altitude at point A is $\frac{\sqrt{2}}{2}$ km ($\approx .707$ km) and at point B, the altitude is $\frac{1}{2}$ km.

Once you found numerous points (x, y) for S(x), you plotted the points and connected them with a smooth curve. Your graph of S(x) should have looked like the following.

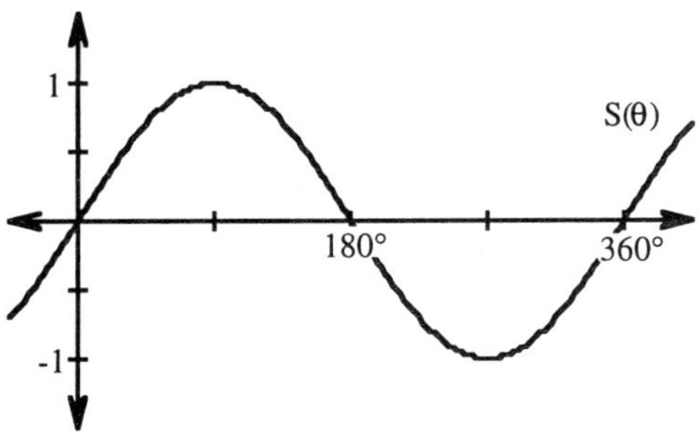

The curve continues cycling up and down in this fashion.

This graph, and graphs similar to this, introduce new terms and attributes to consider. You should be prepared to investigate these and similar functions completely. The domain for the function sine is all real numbers. The range is much more restricted: $-1 \le y \le 1$. It has only one y-intercept but infinitely many x-intercepts!

The new terms to consider (and know!) include **amplitude** (one-half the distance between the highest and lowest value in the range), **cycle** (the part of the graph that is repeated) and **period** (the length of one cycle). For sine (and cosine), the amplitude is 1 and the period is 2π. Try this next example to check your understanding.

1. The sketch of a graph is shown below.

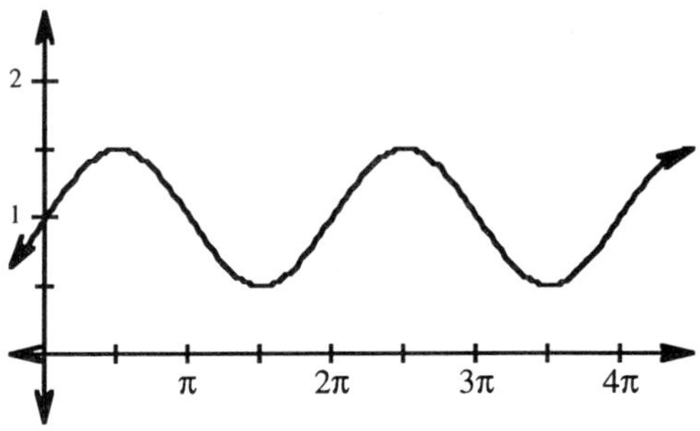

a) Sketch one cycle.

b) What is the domain and range of this graph?

c) What is the period?

d) What is the amplitude?

e) What do you think is the parent graph for this function?

Solution. A cycle is just the shortest piece of the graph that is repeated again and again. In this example, the curve that is repeated is shown at right. Note: it is true that you could say that two of these pieces, one after the other, is also repeated. But, when we are talking about the cycle with trig functions, we want the shortest piece that is repeated.

The domain is all real numbers, and the range is $0.5 \le y \le 1.5$. The period is the length of the cycle. Here, the period is 2π. The amplitude is 0.5 ($1.5 - 0.5 = 1$, $1 \div 2 = 0.5$). What about the parent for this graph? It looks a lot like the graph of sine, just squeezed a little bit and shifted up. The parent is the sine function.

The graph in the last example that looks like the sine function, just squeezed and shifted, is just that: the sine function squeezed and shifted! All of the techniques we have learned to move graphs apply to the graphs of trigonometric functions as well. Before we see just how we do this with trig functions, let's look at an example we did back in our review of Unit 4. Below is Problem 3 from Unit 4, along with the complete solution shown earlier. This is important! Try it again and read through the solution before we try this with trig functions.

3. Suppose $f(x)$ is shown at right.

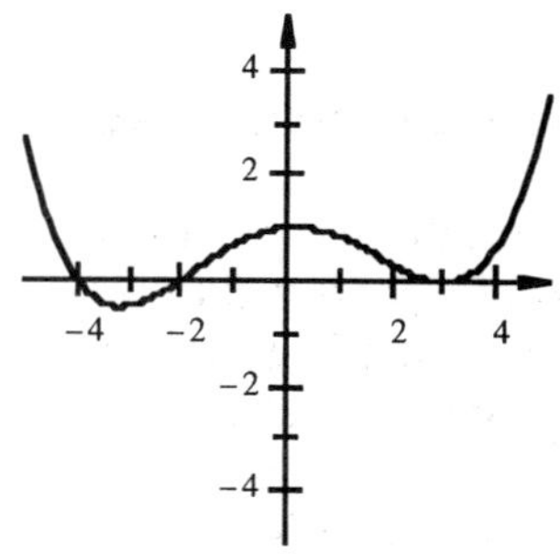

a) What does $f(x) + 3$ look like? Graph it.
b) Graph $f(x) - 2$.
c) Graph $f(x - 1)$.
d) Graph $f(x + 3)$.
e) Graph $3f(x)$.
f) Graph $\frac{1}{2} \cdot f(x)$.

Solution.

Each time we alter the equation slightly, the graph is changed. Even though we have no idea what the equation of this function is, we can still move it around. Think of the big picture when moving graphs. In part (a) we want to find out what $f(x) + 3$ looks like. Remember that $f(x)$ represents the range or y values. Therefore this says "the y values, plus 3" Adding three to all the y values will lift the graph up three units. This is shown at right. Notice that the shape of the graph is exactly as the original, just shifted up three units. Check this by comparing the y-intercepts.

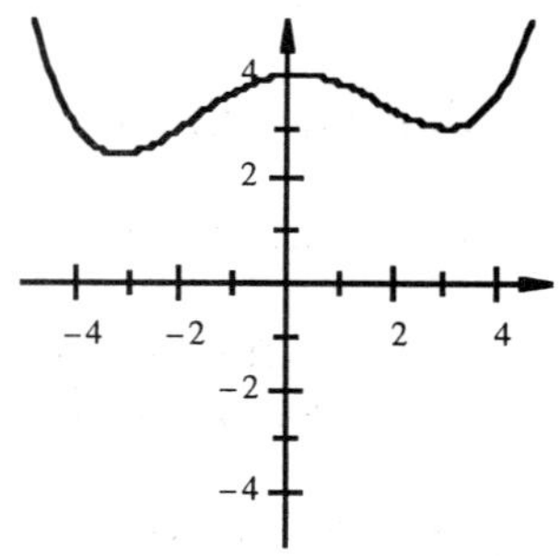

You can probably easily do part (b) now. If $f(x) + 3$ shifts the graph up three units, then $f(x) - 2$ will shift the graph down two units. Again compare the y-intercept in the graph at left to the original. Using the y-intercept or the x-intercepts to help you graph is the smart way to create a graph. These points, as well as a vertex and other key points, are often called **locator points** because, as the name implies, they help you locate the position of the graph.

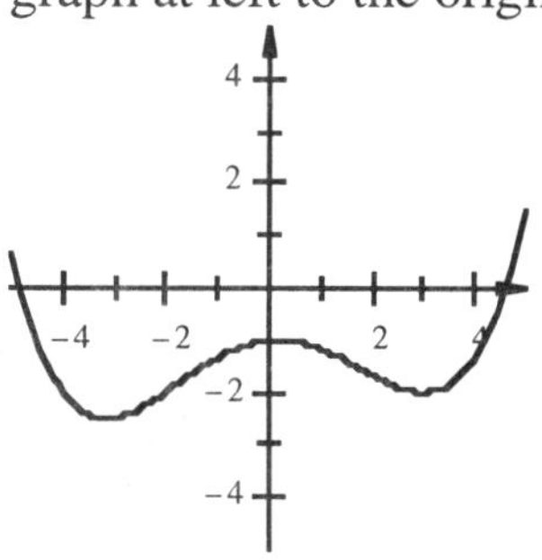

So what happens when the change is made within the parentheses as in parts (c) and (d)? Here we are making the shift with the x coordinates, thus the graph will move left or right. But, be careful! In part (c), the graph of $f(x - 1)$ is shifted to the **right**.

This seems backwards, but it is correct. To help me remember this is I think of the x-intercepts. What is the x-intercept for $y = x - 1$? Solve $x - 1 = 0$. It is + 1, or think of it as one unit to the right. What is the x-intercept for $y = x + 3$? Solve $x + 3 = 0$. It is - 3, or three units to the left. If this doesn't help you remember which way to move, try to develop your own method. Whether you use this idea or develop one of your own, make sure you have the important information in your tool kit.

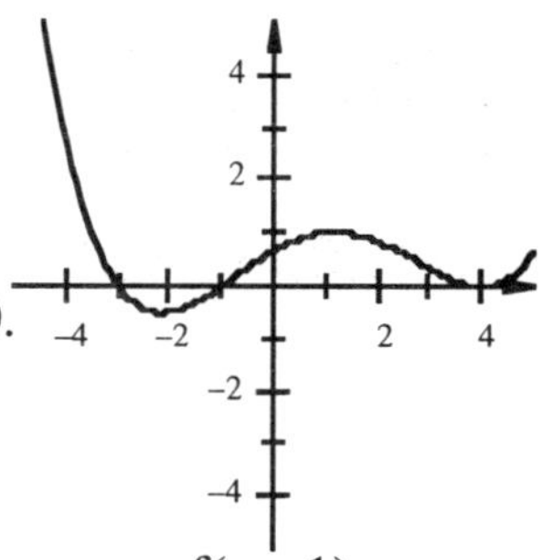

$f(x - 1)$

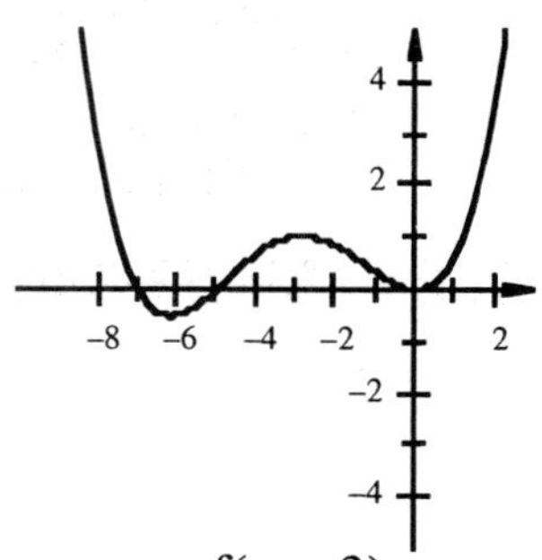

$f(x + 3)$

Multiplying f(x) by a number, as in parts (e) and (f), is what gives students the most trouble. Looking at some key points will help. To begin with, at the x-intercepts $f(x) = 0$, so multiplying by 3 or $\frac{1}{2}$ won't change anything: $3 \cdot 0 = 0$ and $0 \cdot \frac{1}{2} = 0$. Therefore, the x-intercepts do not change at all. What about the y-intercept? In the original graph, the y-intercept is 1, so $f(x) = 1$ there. Multiplying by 3 will raise that point three times as high, to the point (0, 3). Multiplying by $\frac{1}{2}$ changes the y-intercept to $\left(0, \frac{1}{2}\right)$. The larger the constant that you multiply by, the more stretched out the graph becomes. A small number flattens the graph.

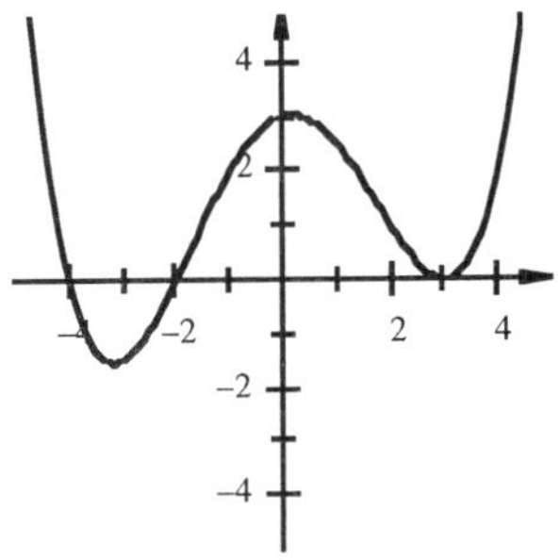

3f(x)

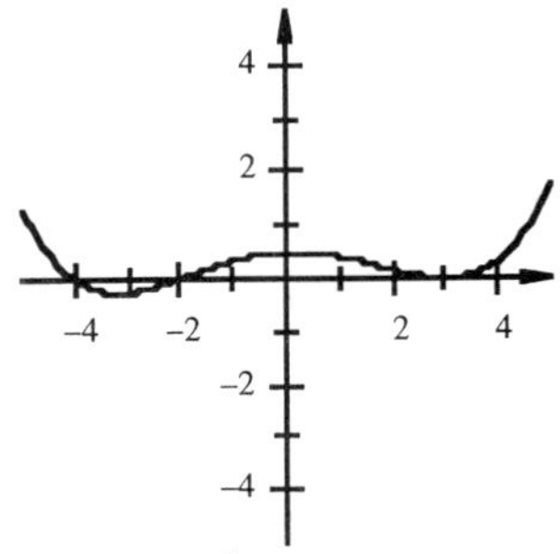

$\frac{1}{2}$ f(x)

Remember doing this problem? You had no idea what the equation (function) might be for this graph, yet you could move it around and change its shape. If you can do this for an unknown function, then you can do it for a trig function. We will use the same ideas we used in problem 3 in the next problem. Try it on your own first.

2. The graph of y = sin x is shown below right. Use it to do the following <u>by hand</u> (i.e. without your calculator!).

Sketch the graph of

a) $y = \sin x + 2$.

b) $y = \sin x - 1$.

c) $y = \sin(x - \pi)$.

d) $y = \sin(x + \frac{\pi}{2})$.

e) $y = 3\sin x$.

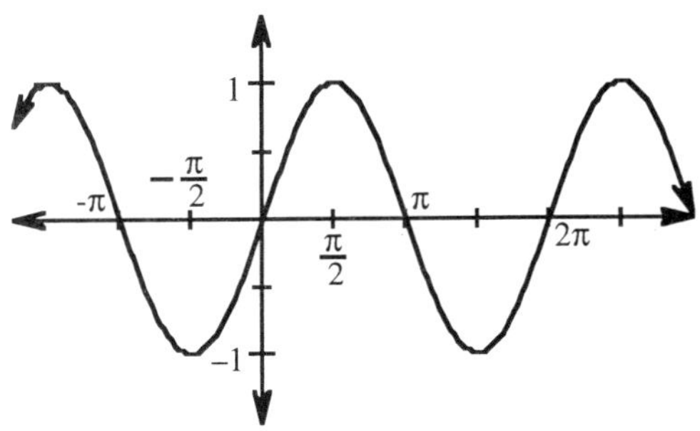

Solution.

It might be helpful to trace this graph so that you can actually move the graph according to the change in the equation. The graphs of parts (a) and (b) shift vertically. Part (a) moves up two units and part (b) moves down one unit. These are drawn on the same axes at right. Notice the shape has not changed at all, but I did change the scale on the y axis to fit both graphs in. Remember to use locator points to help you graph the new function. With trig functions, you have <u>lots</u> of locator points to help you graph. Since there are infinitely many x-intercepts, you can shift many of those. Also, the high and low points (somewhat like vertices) can be shifted too. With the high/low points and the

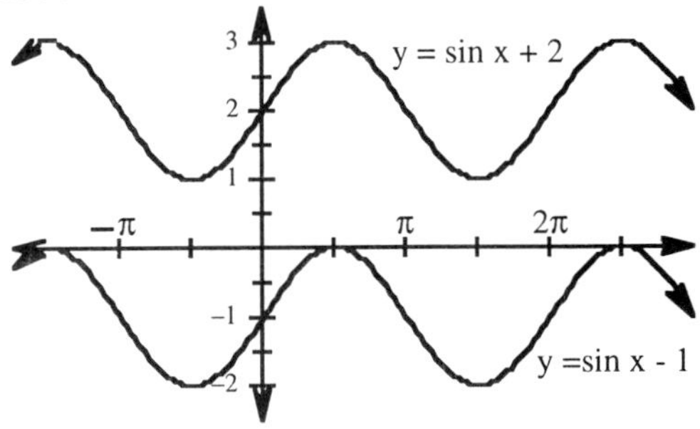

x-intercepts you can easily sketch a graph. Notice also that the amplitude, period, and domain have not changed as well. The range has! What are the new ranges? (For part (a), $1 \le y \le 3$. For part (b),$-2 \le y \le 0$.)

The graphs of parts (c) and (d) will be shifted horizontally. We must be careful when graphing these, not because they are difficult, but because the picture can become muddled or confusing. Since these graphs are cyclic (repeating) sometimes when you start to shift the graph you forget what is going on! Start with just one cycle and shift that. Colored pens or pencils will surely help here. In part (c), the graph is shifted right π units. If I'm using locator points, this means that the high point which happens to be at $x = \frac{\pi}{2}$, will move to the right to the point $x = \frac{3\pi}{2}$. If you did trace the cycle, then you can slide the graph to the right -- not moving it up or down at all -- until the high point is where you want it. That is where the graph will go. Once you have one cycle placed, then you can repeat the cycle to complete the sketch. The completed graph is shown below. I have left the dotted part in so that you can see how the cycle shifted.

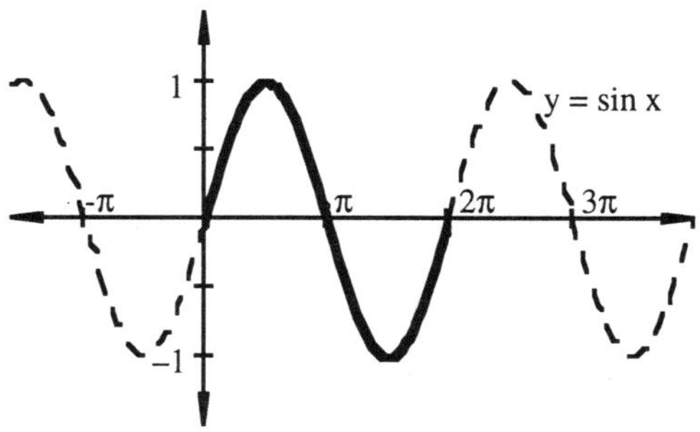

Note that if you are looking at these graphs quickly, you might not notice the difference between this graph and the original! That is why you must do the horizontal shifts carefully. The extra time you take will pay off by avoiding careless errors.

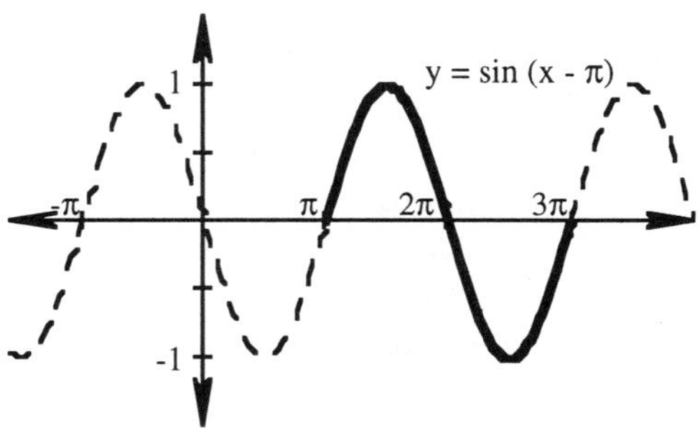

The graph of part (d) is shifted to the left one unit. Again we use the locator points but we must approximate how much one unit is. Make your best educated guess and then label the graph to reflect the shift as one unit. Again it is probably best to trace a cycle and move one cycle first. The completed graph is shown at right. You have to be looking much more closely to see the difference between this graph and y = sin x!

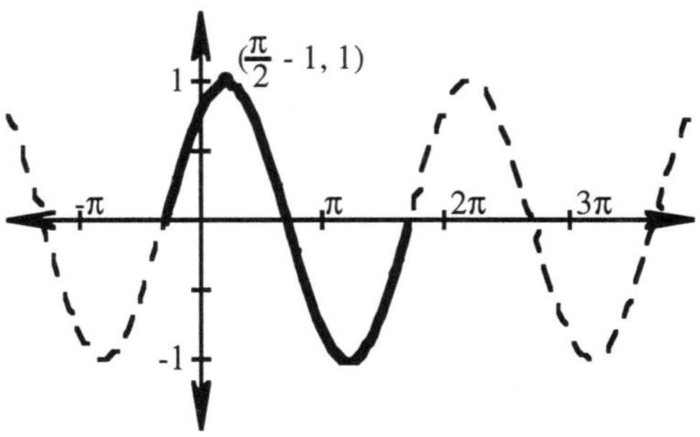

We know that multiplying the function by a constant, as in part (e) stretches or flattens the graph depending on what number we multiply by. Since we are multiplying by 3, this graph will be stretched. Remember: the x-intercepts won't be changed when we do this! The completed graph is shown at right. Notice the scale on the y-axis.

The amplitude of this graph has changed! Now the amplitude is 3, the same as the number we multiplied by! Coincidence?

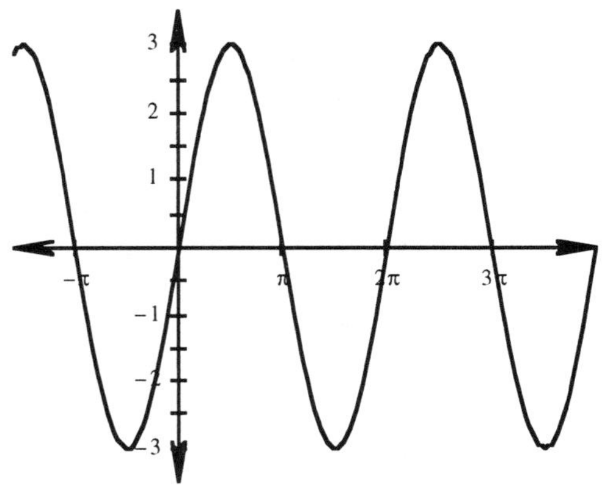

These rules operate in the same manner as they did in earlier units. If you didn't quite understand shifting graphs of functions earlier, think of this unit as another opportunity to master this skill.

There is one other way we can alter the graph. We could have made this change to earlier functions as well. With trig functions, it is often necessary to make this change in the equation: What happens when we multiply x by a constant? Try this next problem.

3. The graph of y = sin x is shown below. Use it to graph the following by hand.

a) $y = \sin(2x)$.

b) $y = \sin(5x)$.

c) $y = \sin\left(\frac{1}{2}x\right)$

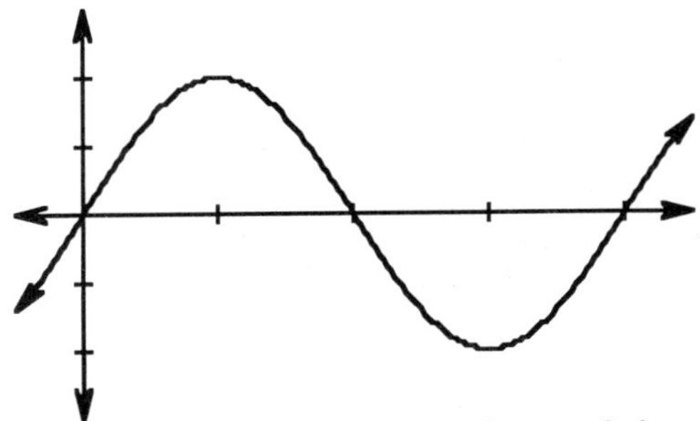

Solution.

We know what happens when we multiply the function by a constant as in $y = 3\sin x$, so what happens when we multiply the variable by a constant? Let's examine a couple of locator points to see what happens with those. The point $x = 0$ is an x-intercept for $y = \sin x$ because $\sin 0 = 0$. In the equation $y = \sin(2x)$, if $x = 0$ we have $y = \sin(2 \cdot 0) = \sin 0 = 0$, so $x = 0$ still is an x-intercept. Does this change the graph at all? It does and here's how. Since we are now taking $\sin(2x)$, we would only need half the original points to get the same graph. For instance, suppose the point (a, b) is a point on the graph of $y = \sin x$. This means that $b = \sin a$. In the new equation, b would not necessarily equal $\sin(2a)$. But, if we substituted $\frac{1}{2}a$ we would have $\sin(2 \cdot \frac{1}{2}a) = \sin a = b$, so $\left(\frac{1}{2}a, b\right)$ is a point on the graph. What this does is shrink the period and the length of the cycle is cut in half.

The graph of $y = \sin x$ is drawn with a dashed curve while the graph of $y = \sin(2x)$ is drawn with a solid curve. Notice that in the span that it takes $y = \sin x$ to make one cycle, $y = \sin(2x)$ has made two cycles. What do you think happens with $y = \sin(5x)$?

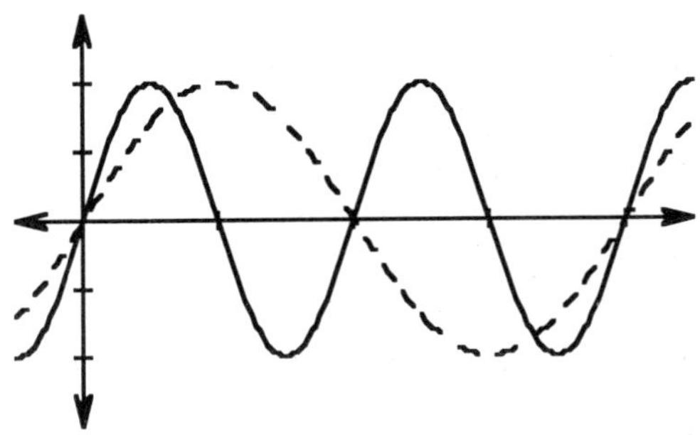

If you said that the graph will do five cycles in the space of one cycle of $y = \sin x$, you are correct. Look at the graph at right closely ($y = \sin x$ is graphed with a dashed line again). Starting at $x = 0$, count the number of cycles to be convinced that there are five of them in the same space that $y = \sin x$ completes one.

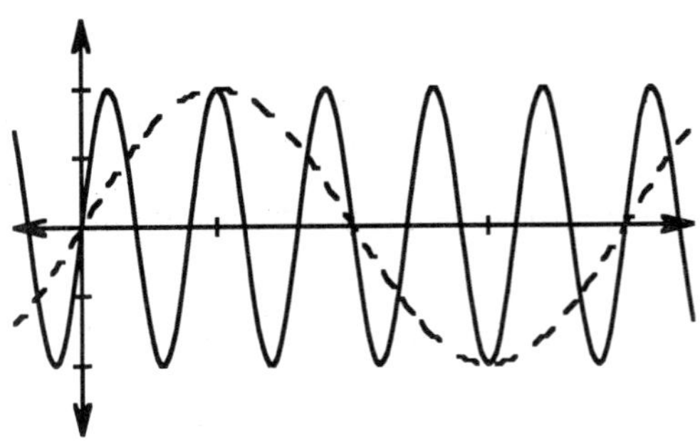

Last but not least, what happens with $y = \sin\left(\frac{1}{2}x\right)$? Use the very same reasoning. If $y = \sin(5x)$ does five cycles in place of one, then $y = \sin\left(\frac{1}{2}x\right)$ will do one-half a cycle in place of one. Look at the graph at right. We don't even see a complete graph of $y = \sin\left(\frac{1}{2}x\right)$ when we see a complete graph of $y = \sin x$.

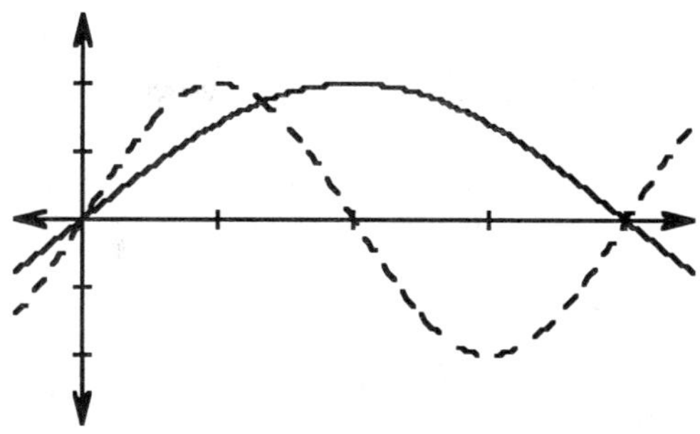

In general, $y = \sin(nx)$ changes the graph by drawing n cycles in the normal span of one.

Before we begin modeling data with trig functions, we need to review one other thing.

There is more than one way to measure angles, you might have noticed. WE have measured central angles in degrees, with one complete revolution being 360°. But the more common way to measure angles, particularly when discussing functions, is with **radians**. With radian measure, one complete revolution measures 2π radians: the distance traveled along the circumference in that one revolution. Let's look at this closer.

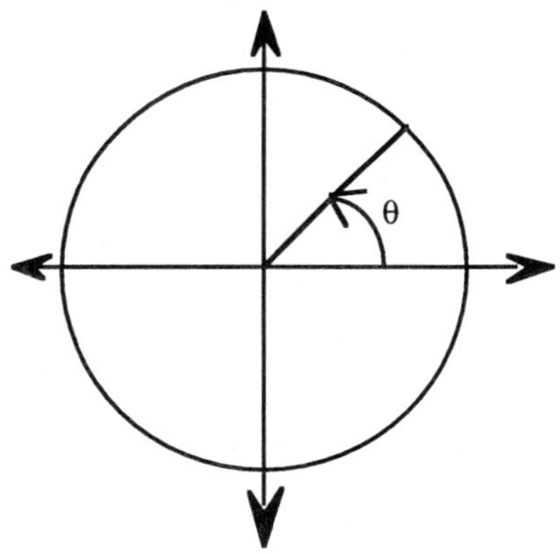

Consider a unit circle with an angle being swept out in standard position (that is, starting at the x axis and moving counter-clockwise). In one complete revolution, the angle will have swept out 360° (that's degrees!). In the same action, the angle also has swept out 2π radians. Knowing that a complete revolution, which in degrees is 360°, is also 2π in radians, we can figure out any other angle measure. (This is somewhat like converting temperatures from Fahrenheit to Celsius.) A 180° angle is half-way around the circle so in radians that would be $\frac{2\pi}{2} = \pi$ radians. A 90° angle is one-fourth of the circle so $90° = \frac{2\pi}{4}$ radians $= \frac{\pi}{2}$ radians. What about a 270° angle? That's three-fourths of a complete revolution so $270° = \frac{3}{4}(2\pi)$ radians $= \frac{3\pi}{2}$ radians.

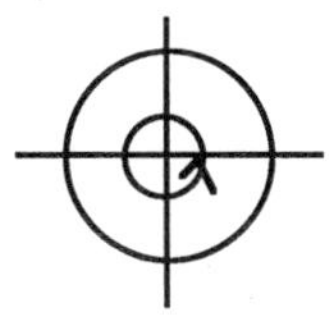

360° = 2π radians

What do you do when you have a 20° angle or some other angle where it is not immediately obvious how much of a revolution you might have? Just remember that 360° = 2π radians. One way to convert 20° to radians is to use ratios:

$$\frac{20°}{360°} = \frac{x}{2\pi \text{ radians}} \;\text{-->}\; 360x = 40\pi \text{ radians} \;\text{-->}\; x = \frac{40\pi}{360} \text{ radians} = \frac{\pi}{9} \text{ radians}$$

Another way is to find what one degree equals or what one radian equals:

One degree: $360° = 2\pi$ radians, so $1° = \frac{2\pi}{360°}$ radians $= \frac{\pi}{180°} \approx 0.017$ radians

One radian: $360° = 2\pi$ radians, so $1 \text{ radian} = \frac{360°}{2\pi} = \frac{180°}{\pi} \approx 57.3°$

When you graph a standard trig function using radians, the length of one cycle is 0 to 2π radians. With degrees, it is 0 to 360°.

See if you can draw each of the following angles in standard position when the measures are given in radians.

4. For each of the following, draw a unit circle. Then draw the given angle in standard position.

a) $\Theta = \frac{\pi}{4}$ **b)** $\Theta = \frac{\pi}{6}$ **c)** $\Theta = \frac{4\pi}{3}$ **d)** $\Theta = \frac{5\pi}{6}$

Solution. You may find that you can draw these without converting them to degrees. That is great! You can do that if you remember that the whole circle is 2π radians and you figure out what fraction of that you have with Θ. In each picture below I also included the degree measure of the angle as an added check.

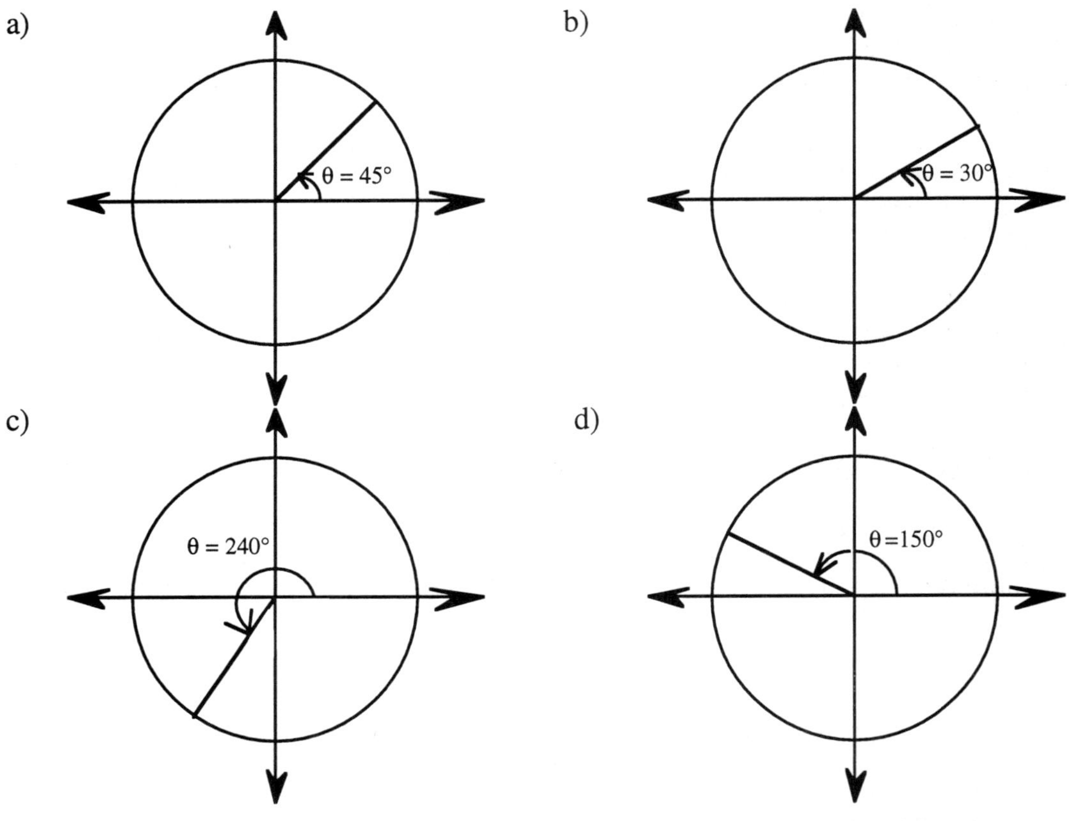

Radian measure is used on almost all of the application (modeling) problems. If you are planning to take any more math classes--calculus for instance--you will find radians to be the angle measure of choice. Try to start thinking in terms of the unit circle (the circumference of a unit circle is 2π) and what fractional part of the circle you have and radians will start to seem natural to you. Note: be sure you know how to switch your calculator from degrees to radians and back again.

Now we are ready to try an application. If you have read through everything we have done so far, then you are ready to try it on your own.

5. **For the Fourth of July parade, Tess decorated her tricycle with streamers and balloons. She stuck one balloon on the outside rim of one of her back tires. The balloon starts at ground level. As she rides, the height of balloon rises up and down, sinusoidally. The diameter of her tire is 10 inches.**

a) **Sketch a graph showing the height of the balloon above the ground as Tess rolls along.**

b) **What is the period of this graph?**

c) **Write the particular equation of this function.**

d) **Use your equation to predict the height of the balloon when Tess has traveled 42 inches.**

Solution.

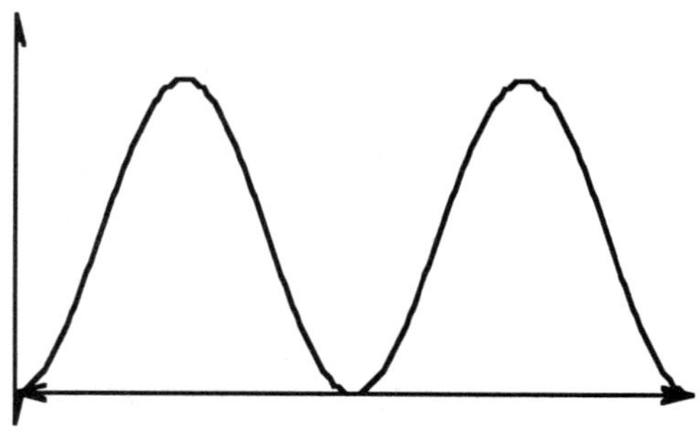

This is somewhat similar to the Ferris Wheel problem, CF-21 in Unit 7. The balloon is rising up and down just as a sine or cosine curve rises up and down. A simple sketch is shown at right.

The balloon begins next to the ground and as the tricycle rolls, the balloon rises to the top of the wheel, then comes back down. How high does this balloon get? How far from he ground? Here, the x-axis represents the ground. The balloon is at its highest point when it is at the top of the wheel, a distance of one wheel's height from the ground. But one wheel height is just the diameter of the wheel, 10 inches. So now we know that the distance from the highest point to the lowest point is 10. From this we can figure out the amplitude because it is half of this distance. Therefore the amplitude is 5.

What about the period of this curve? Think about the problem. The balloon starts at ground level, rises as the wheel rolls and comes down again to the ground. What has happened when the balloon returns to the ground? The wheel has made one complete revolution. How far has the wheel traveled then? It has traveled the distance of one circumference. The circumference of a circle with diameter 10 inches is 10π inches. Therefore the period of this graph is 10π. This is <u>not</u> in degrees so we must be working in radians.

So how are we going to get the particular equation for this graph? At this point, we need to make some decisions. You might have noticed how similar the graphs of sine and cosine are. In fact, one is just the other shifted $\frac{\pi}{2}$ radians (or 90°). At this point, we need to decide if we want to use sine or cosine to model this data. Either one will work but the answers will look different. Since the graph starts at the low point, I'm going to use cosine. Yes, cosine starts at the high point but all we need to do is multiply by a negative to flip the graph over. (Remember all of the techniques we have learned for moving graphs!) I also know the amplitude of the graph is 5. So I know that the equation looks like this: $y = -5\cos(ax) + b$. I know there is some number a, not equal to one, that multiplies x. If a = 1, then the period would be the standard 2π, which it is not. Since the period is 10π, we must find out what a is to make the period change. Remember that when we graphed $y = \sin(2x)$, the period was $\frac{2\pi}{2} = \pi$. When we graphed $y = \sin(5x)$, the period was $\frac{2\pi}{5}$. When we graphed $y = \sin\left(\frac{1}{2}x\right)$, the period was $\frac{2\pi}{0.5} = 4\pi$. Now we want a such that $\frac{2\pi}{a} = 10\pi$. Solving this for a gives $a = \frac{1}{5}$. (Check that yourself!) Now the equation is $y = -5\cos\left(\frac{1}{5}x\right) + b$. We need b to be a positive number. If b were zero, as in the standard equation, the graph would rise above and below the x-axis. Our graph comes down and is tangent to the x-axis and does not pass below it. It is shifted up above the axis five units. Therefore the equation is $y = -5\cos\left(\frac{1}{5}x\right) + 5$. Check this by graphing it.

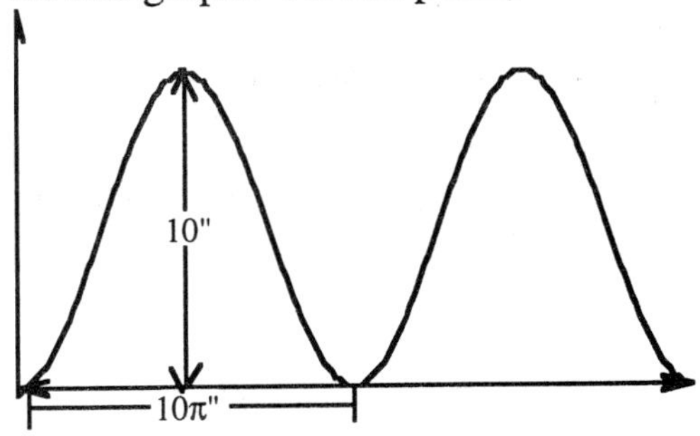

Note: if you decided to use the sine function for this data, you must realize that the graph is shifted to the right $\frac{10\pi}{4}$ units. One equation that gives this graph is

$y = 5\sin\left(\frac{1}{5}x - \frac{10\pi}{4}\right) + 5$. Remember to graph the equation you wrote to check if it graphs the same curve. There are many different equations that will give this graph so you don't have to get one of the two I've written.

To find the height of the balloon after Tess rides 42 inches, we substitute 42 for x in the equation.

$$y = -5\cos\left(\frac{1}{5}42\right) + 5 = -5\cos(8.4) + 5 \approx 7.596 \text{ inches}$$

If you can't get this answer, make sure your calculator is in radians!

You may not have anticipated that Unit 8, a unit on trigonometry, could be a good review of Unit 4, Parent Graphs, but it is. This unit lets you review and practice all of the skills you learned about altering graphs and moving them around. At this point in the year you should be getting comfortable doing that.

I also want to remind you to keep your tool kit up to date and to make sure you can use your calculator correctly. As I mentioned earlier, you need to be especially careful about knowing when and how to change your calculator from degree mode to radians and back again. If you find that you think you are doing a problem correctly and just can't seem to get the graph or answer you are hoping for, check the mode. Maybe you just need to switch to radians (or degrees!) to get the correct answer.

Unit 8
More To Try

Examine each graph below. For each one, draw a sketch of one cycle, give the amplitude and the period.

1.

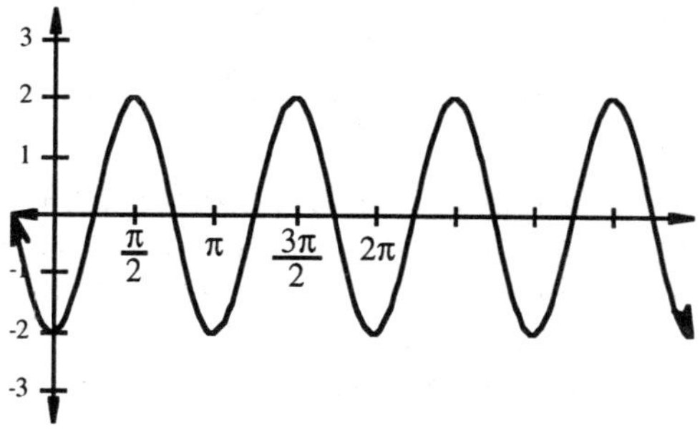

2

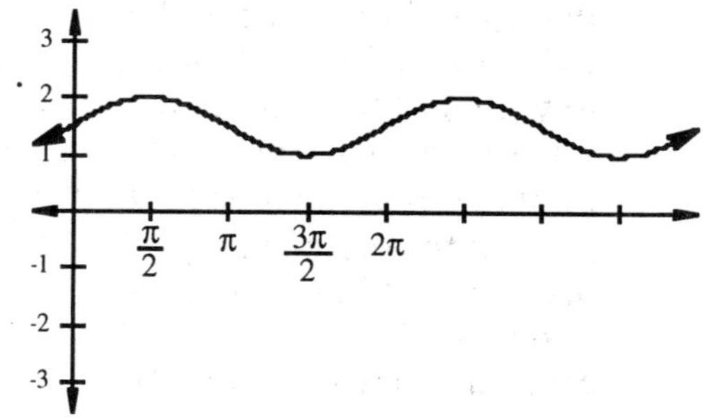

3.

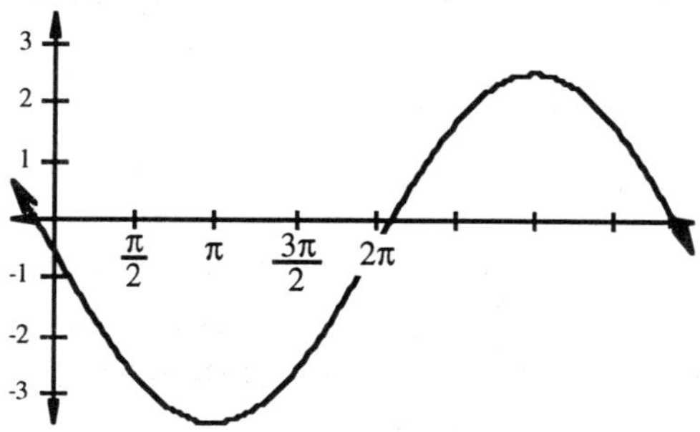

4

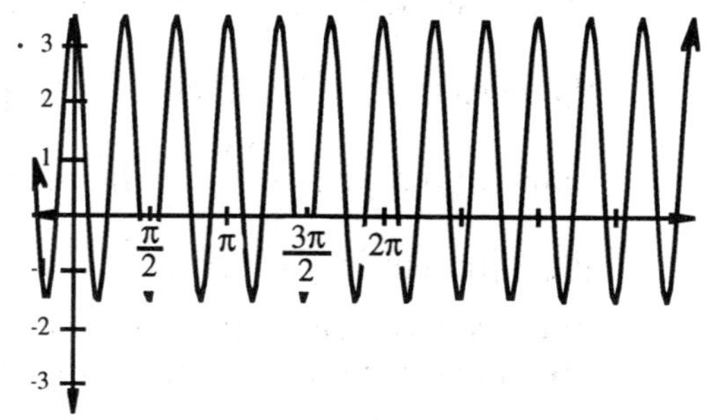

For each equation listed below, state the amplitude and period for each.

5. $y = 2\cos(3x) + 7$

6. $y = \frac{1}{2}\sin x - 6$

7. $f(x) = -3\sin(4x)$

8. $y = \sin\left(\frac{1}{3}x\right) + 3.5$

9. $f(x) = -\cos x + 2\pi$

10. $f(x) = 5\cos(x - 1) - \frac{1}{4}$

Draw a picture of an angle in standard position using the measures below. Then convert the angle measure from degrees to radians or radians to degrees.

11. $\theta = 70°$

12. $\theta = 110°$

13. $\theta = \frac{\pi}{5}$

14. $\theta = \frac{7\pi}{4}$

15. $\theta = 330°$

16. $\theta = \frac{4\pi}{3}$

17. $\theta = -\frac{\pi}{3}$

18. $\theta = -90°$

Here are the graphs of $y = \sin x$ and $y = \cos x$.

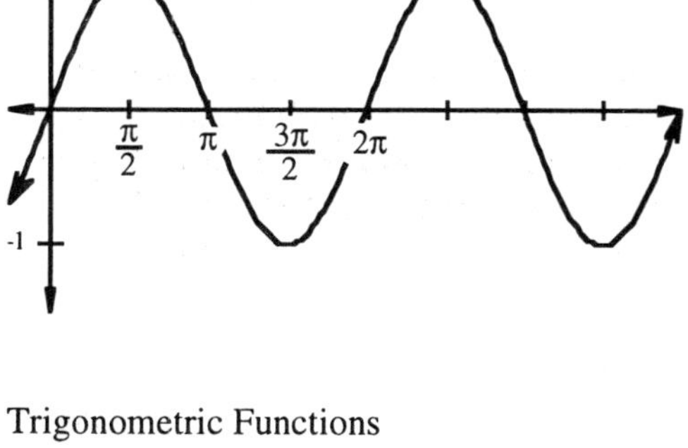

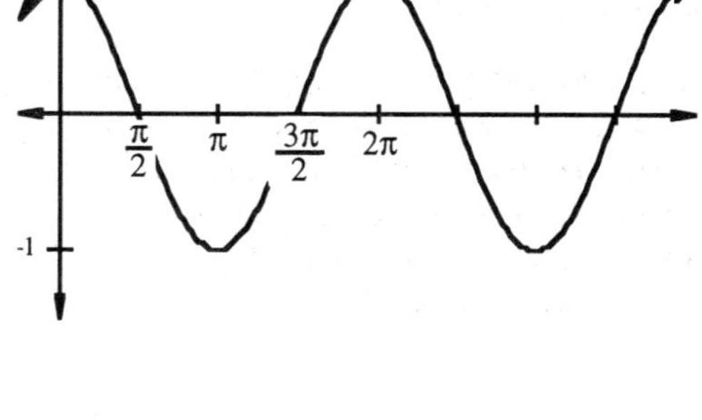

Use these equations to sketch the graphs of each of the following by hand. Use your graphing calculator to check your answer.

19. $y = -2\sin(x + \pi)$

20. $f(x) = 0.5\sin(3x)$

21. $f(x) = \cos\left(4\left(x - \frac{\pi}{4}\right)\right)$

22. $y = 3\cos\left(x + \frac{\pi}{4}\right) + 3$

23. $f(x) = 7\sin\left(\frac{1}{4}x\right) - 3$

24. A wooden water wheel is next to an old stone mill. The water wheel dips down two feet below the surface of the water and at its highest point is 18 feet above the water. A snail attaches to the edge of the wheel when the wheel is at its lowest point and rides the wheel as it goes round and around. As time passes, the snail rises up and down and in fact, the height of the snail above the surface of the water varies sinusoidally with time. Use this information to write the particular equation which gives the height of the snail over time.

25. To keep baby Cristina entertained, her mother often puts her in a Johnny Jump Up. It is a seat on the end of a strong spring which attaches in a doorway. When Mom puts Cristina in, she notices that the seat drops to just 8 inches above the floor. One and a half seconds later (1.5 seconds), the seat reaches its highest point of 20 inches above the ground. The seat continues to bounce up and down as time passes. Use this information to write the particular equation which gives the height of baby Cristina's Johnny Jump Up seat over time. (Note: You can start the graph at the point where the seat is at its lowest.)

SAT PREP

1. If $b + 4 = 11$, then $(b - 2)^2 =$

   a) 16 b) 25 c) 36 d) 49 e) 64

2. Let P and Q represent digits in the addition problem shown at right. What must the digit Q be?

$$\begin{array}{r} 25P \\ +\ P4 \\ \hline 32Q \end{array}$$

   a) 0 b) 1 c) 2 d) 3 e) 4

3. If $3^4 = 9^x$, then $x =$

   a) 2 b) 3 c) 5 d) 8 e) 10

4. When a positive number n is divided by 7 the remainder is 6. Which of the following expressions will yield a remainder of 1 when it is divided by 7?

   a) n + 1 b) n + 2 c) n + 3 d) n + 4 e) n + 5

5. How many 4-digit numbers have the thousands digit equal to 2 and the units digit equal to 7?

   a) 100 b) 199 c) 200 d) 500 e) 1000

UNIT 9

The Search for Sunken Treasure

MORE TRIANGLES AND TRIGONOMETRY

UNIT 9
MORE TRIANGLES AND TRIGONOMETRY

If you thought that Unit 8 did not seem like the trigonometry you saw in earlier math classes, then this unit might seem more familiar. We go back to triangles in this unit, finding new ways to calculate the sides and angles of triangles. We also learn "officially" how to "undo" trig functions, and we get three new trig functions which are closely related to the three we already know.

We begin by trying a problem requiring the Law of Sines. This law is derived in ST-2 and stated formally in the box following that problem. Read that box and problem again before trying this next problem.

1. In ΔABC, $m\angle A = 36°$, $m\angle B = 68°$, and $b = 8$. Find a and c.

Solution.

The first thing to do with a problem like this is draw a picture! We must see what we have and what we are trying to get. Fill in every piece of information we have about the triangle. In this case, we have the measures of two angles and the length of one non-included side. If this were a right triangle then we could solve with the trig we learned in Math 2. Checking this option we find that the measure of the third angle is 76° (180° - 36° - 68°). Now we can use the Law of Sines to solve for a and c in this triangle.

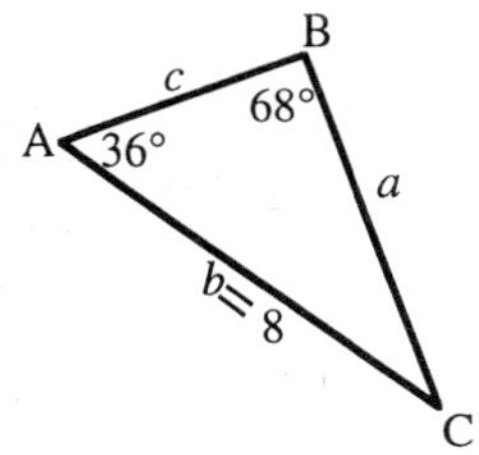

The Law of Sines states that $\frac{\sin A}{a} = \frac{\sin B}{b} = \frac{\sin C}{c}$

where A, B, and C are the angle measures, and a, b, and c are the lengths of the sides opposite those angles. To get started, look at the picture carefully. You need to start with an angle and the side opposite that angle. In this one, since b is the only side we know, we will start with $\angle B$ and b. Let's use it to find a first.

$$\frac{\sin B}{b} = \frac{\sin A}{a}$$
$$\frac{\sin 68°}{8} = \frac{\sin 36°}{a}$$
$$a(\sin 68°) = 8(\sin 36°)$$
$$a = \frac{8\sin 36°}{\sin 68°}, \quad a \approx 5.07$$

Note: Did you get $a \approx 8.84$ no matter how often you check your work? Check the mode for your calculator. This problem is in degrees, not radians.

Now we can use the Law of Sines again to find c. Remember that $m\angle C = 76°$.

$$\frac{\sin B}{b} = \frac{\sin C}{c}$$
$$\frac{\sin 68°}{8} = \frac{\sin 76°}{c}$$
$$c(\sin 68°) = 8(\sin 76°)$$
$$c = \frac{8\sin 76°}{\sin 68°}, \quad c \approx 8.37$$

With the Law of Sines you are not restricted to finding the sides and angles of right triangles. But the Law of Sines does not work for every triangle. You must know at least one pair of an angle and its side opposite the angle. For instance, in the picture below, try to find an unknown side or angle by using the Law of Sines. No matter how hard you try, you will end up with one equation, but two unknowns in the equation. Try it! You just can't solve it. So what do you do with a triangle like this? You use the Law of Cosines. Read ST-42, ST-54, ST-55, and the box after ST-56 before trying this next problem.

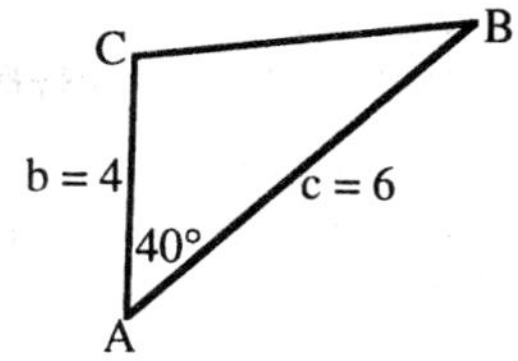

2. The lengths of two sides and the measure of the included angle are given as shown in the triangle at right. Find the measures of the other two angles and the length of the third side.

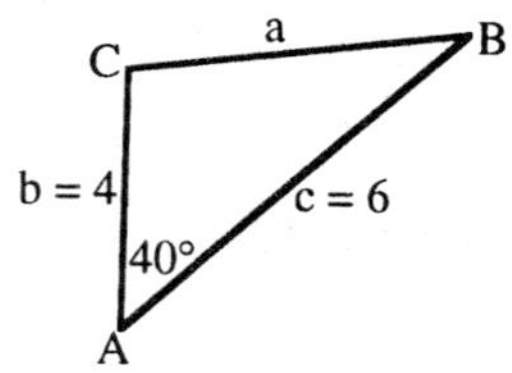

Solution.

Don't worry about not knowing when to use Law of Sines and when to use Law of Cosines. You can always start to use one and, as we saw above, if it doesn't work, you will find out soon enough. We already realized that the Law of Sines doesn't help us here, so let's use the Law of Cosines.

If ΔABC has angles A, B, and C and sides opposite those angles of a, b, and c, respectively, then the Law of Cosines states:

$$a^2 = b^2 + c^2 - 2bc \cos A$$

and is also true for the other sides as well.

$$b^2 = a^2 + c^2 - 2ac \cos B$$
$$c^2 = a^2 + b^2 - 2ab \cos C$$

Let's substitute the values we know to find the length of side a.

$$a^2 = b^2 + c^2 - 2bc \cos A$$
$$a^2 = 4^2 + 6^2 - 2(4)(6) \cos 40°$$
$$a^2 = 16 + 36 - 48 \cos 40°$$
$$a^2 \approx 15.23$$
$$a \approx 3.90$$

Remember that in solving a second degree equation there should be two answers, a positive one and a negative one. We don't need to worry about the negative one because a is the length of a side of a triangle; it can't be negative!

Now that we know a we can find the measures of the two missing angles. Here, we don't have to use Law of Cosines again. Now that we have an angle and the side opposite it, we can use Law of Sines.

$$\frac{\sin A}{a} = \frac{\sin B}{b} \text{ --> } \frac{\sin 40°}{3.9} = \frac{\sin B}{4} \text{ --> } \frac{4\sin 40°}{3.9} = \sin B \text{ --> } m\angle B \approx 41.24°$$

To find the measure of $\angle C$ we can just use the fact that the three angles sum to 180°.

$$m\angle C = 180° - 40° - 41.24° = 98.76°$$

Some students mistakenly think they only use Law of Cosines when they want to find the length of a side. The Law of Cosines is also used to find angle measures, as is the case in this next problem.

3. In ΔABC, a = 2, b = 3, and c = 4. Find the measures of each of the angles.

Solution.
Let's draw a picture to help us keep track of what is going on. In this example we don't know a single angle measure, we just know the lengths of all three sides. Be sure to label the sides in the proper position, opposite the correct angle. Try to use Law of Sines. What happens? You get that one equation with two unknowns again. Let's try the Law of Cosines, using ∠A.

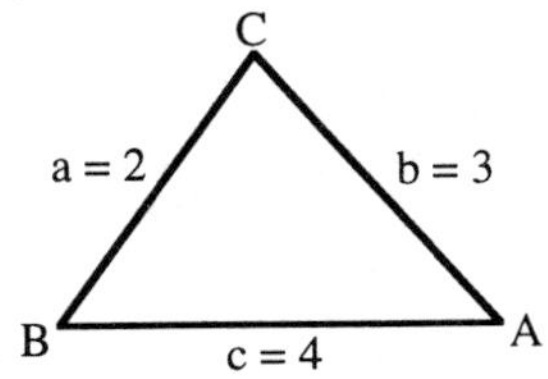

$$\begin{aligned} a^2 &= b^2 + c^2 - 2bc \cos A \\ 2^2 &= 3^2 + 4^2 - 2(3)(4) \cos A \\ 4 &= 9 + 16 - 24 \cos A \\ 4 &= 25 - 24 \cos A \\ -21 &= -24 \cos A \\ \frac{7}{8} &= \cos A \\ m\angle A &\approx 28.96° \end{aligned}$$

Note: Watch your order of operations on the third line down. That's where many students make mistakes.

We need to do the same thing one more time to find the measure of another angle, or, we can use the Law of Sines to find the measure of another angle. Usually, I would choose to use Law of Sines. But, right now we need the practice with the Law of Cosines so I'll use it one more time.

$$\begin{aligned} b^2 &= a^2 + c^2 - 2ac \cos B \\ 3^2 &= 2^2 + 4^2 - 2(2)(4) \cos B \\ 9 &= 4 + 16 - 16 \cos B \\ 9 &= 20 - 16 \cos B \\ -11 &= -16 \cos B \\ \frac{11}{16} &= \cos B \\ m\angle B &\approx 46.57° \end{aligned}$$

We can easily find the measure of angle C now.

$$\begin{aligned} m\angle C &\approx 180° - 28.96° - 46.57° \\ &\approx 104.45° \end{aligned}$$

Most of the applications you will find are exactly the same as what we have just done <u>except</u> that the triangle is describe in some "real" world setting, the shape of a patio, surveyor's measurements for a plot of land, etc. With these applications, be sure you draw a picture and label it correctly. That will be key to solving the problem successfully.

One application that is slightly different is that you can use the Law of Sines to help you find the area of triangles when you don't know the length of a height of a triangle. As you probably remember, the area of a triangle is found by

$$A = \frac{1}{2}bh$$

where b is the side of the triangle and h is the length of the height perpendicular to the known side. But as you'll see in this next example, we don't have that information.

4. Find the area of the triangle shown at right.

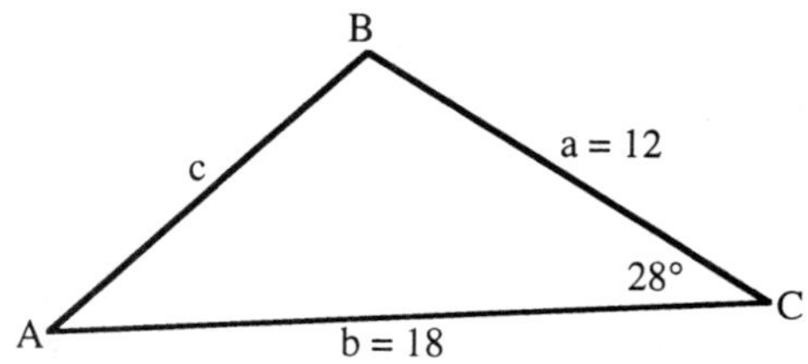

Solution.
If we had the length of a height from the point B down to the side b, then we could easily find the area. But we don't. But in problems ST-14 through ST-18, you developed a formula to find the area using the Law of Sines. If you are not the sort of student who easily remembers formulas (or even if you are!), this is one of those formulas you can easily come up with when you need it.

Draw in a height from B, perpendicular to the base b. It made a right triangle! We can use trig to find the height of the triangle.

$$\sin 28° = \frac{h}{12}$$

$$12 \sin 28° = h, \quad \text{so } h \approx 5.63$$

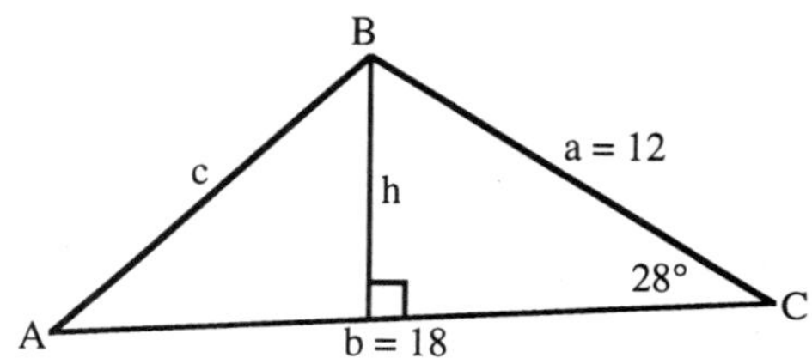

Now that we know h we can find the area.

$$A = \frac{1}{2}(18)(5.63)$$

$$A \approx 50.67 \text{ square units.}$$

This solution was pretty straightforward. If you like the formula, $A = \frac{1}{2}bc \sin A$, feel free to use it. But be sure to have it in your tool kit correctly and with an example nearby.

There is, unfortunately, a time when solving for unknown parts of a triangle leads to an unclear case. This is known as the **Ambiguous Case** and it happens when we are given the lengths of two sides and the <u>not</u>-included angle. This pattern, you might recall, is denoted by SSA, and it is the same pattern that in geometry could not guarantee us congruent triangles. (Just as a reminder, the triangle in problem 1 exhibited the AAS pattern, the triangle in problem 2 exhibited the SAS pattern and the triangle in problem 3 exhibited the SSS pattern.)

If I told you to draw a triangle with sides 6, 7 and 8 inches, and I drew a triangle with the same lengths for sides, our triangles would be congruent. That is because the triangles have all three sides congruent, exhibiting the SSS congruence pattern we learned in Math 2. If we tried this same experiment with two side lengths and the measure of the included angle congruent, we would get congruent triangle again because of the SAS congruence pattern. However, if I told you the lengths of two sides and the measure of the <u>non</u>-included angle, there would be no guarantee that our two triangles would be congruent. The SSA pattern doesn't <u>guarantee</u> congruent triangles. Given this pattern it is possible to draw two different triangle with those parts.

In the figure at right, the two triangles show the SSA pattern and these triangles are definitely not congruent! Let's try to find the missing parts of the triangle(s) to see what happens.

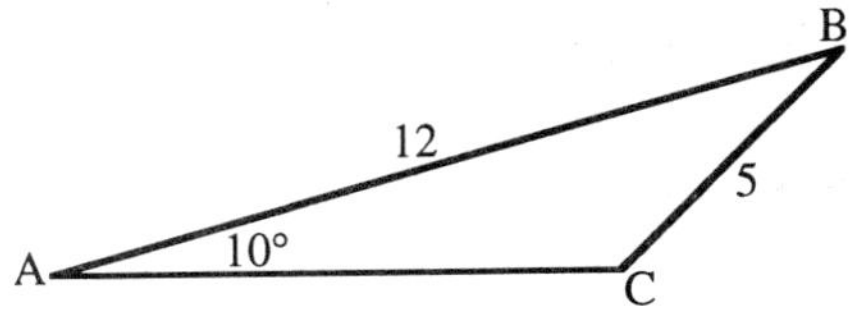

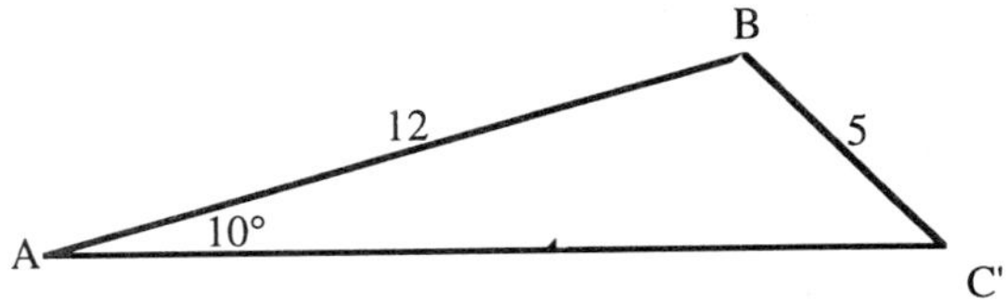

$$\frac{\sin A}{a} = \frac{\sin C}{c}$$
$$\frac{\sin 10°}{5} = \frac{\sin C}{12}$$
$$\frac{12\sin 10°}{5} = \sin C$$
$$\sin C \approx .4168$$
$$m\angle C \approx 24.63°$$

But wait! What's wrong with that? If you look at the top picture you might notice that $\angle C$ is obtuse. How can it equal 24.63°? We must have found the measure of C' in the bottom figure. So what is $m\angle C$ in the top figure?

To see what happened, let's push these two triangles together and do some testing. The $\Delta BCC'$ is isosceles since it has two sides of length 5. What does this tell us about $\angle 1$ and $\angle 2$? They must be congruent (Base angles of an isosceles triangle are congruent.) Since $\angle 1$ and $\angle 3$ form a straight angle, $m\angle 1 + m\angle 3 = 180°$. But since $m\angle 1 = m\angle 2$ we can also write $m\angle 2 + m\angle 3 = 180°$. If $m\angle 2 = 24.63°$, then $m\angle 3 = 155.37°$. Now, look back up to where we used the Law of Sines to find the measure of angle C. We have on the second to last line, $\sin C \approx 0.4168$. From that we calculated the measure of angle C. Try taking sine of 155.37°. What do you get? Approximately 0.4168. There are two possible answers. And since we are using an unclear congruence property, SSA, we cannot be sure which one is the correct answer. Both are reasonable.

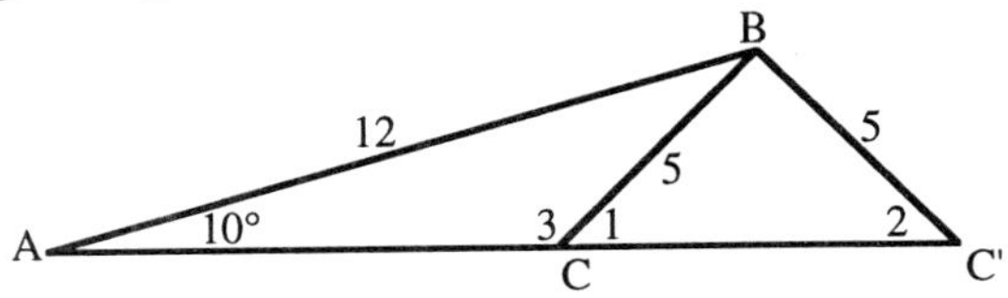

As you might have discovered, or your teacher might have pointed out, solving some trigonometric equations gives multiple answer. Remember the graphs are cyclic. To make this clearer, suppose you wanted to solve $\sin C = 0.4168$ by graphing $y = \sin x$ and $y = 0.4168$. How many times will these two graphs cross? If you are not sure, try it. There are infinitely many solutions. Only two of them are reasonable however since we are working within the confines of a triangle. Another solution is 384.63° but that would never be the measure of an angle in any triangle we have ever seen! We only use the values that make sense.

The moral of this is that whenever you are trying to find the parts of triangles, make sure you have one and only one triangle first. Make sure that the triangle inequality holds and that the parts illustrate one of the congruence patterns that are correct.

In solving such equations we are "undoing" trig functions and most of you have no trouble using the correct keys on your calculator to do this. To solve $\sin x = 0.5$, for instance, we would compute $\sin^{-1} 0.5$ with the calculator. Sin^{-1} is the inverse of sine, also known as arcsin, and there are similar inverses for the other trig functions. The inverse of cosine is $\cos^{-1}$ or arccos and the inverse of tangent is $\tan^{-1}$ or arctan. Are these inverses functions? They are only functions if they are restricted. Read ST-113 through ST-115 before trying these next problems.

5. Solve the following equations for x in degrees.

a) $4\cos x + 5 = 7$ **b)** $(\sin x + 1)(\sqrt{2}\cos x - 1) = 0$

c) $\frac{1}{2}\sin x = 4$

Solution.

Don't be concerned about solving these equations just because they have a trig function in them. You solve them the same way you solved other equations with one change in the last step: you must "undo" the trig function.

Let's solve the equation in part (a).

$$4\cos x + 5 = 7$$
$$4\cos x = 2$$
$$\cos x = \frac{1}{2}$$

Now we simply undo the cosine.

$$x = \cos^{-1} 0.5$$
$$x = 60°$$

Note: there are other solutions, namely x = 300° will work. Check with your teacher to see if you should list more than one solution for these equations.

Part (b) is also done just as we did earlier problems. Here we use the Zero Product Property.

$$(\sin x + 1)(\sqrt{2}\cos x - 1) = 0$$

$$\sin x + 1 = 0 \quad \text{or} \quad \sqrt{2}\cos x - 1 = 0$$
$$\sin x = -1 \qquad \sqrt{2}\cos x = 1$$
$$x = \sin^{-1}(-1) \qquad \cos x = \frac{1}{\sqrt{2}}$$
$$x = 270° \text{ or } -90° \qquad x = \cos^{-1}\left(\frac{1}{\sqrt{2}}\right)$$
$$x = 45° \text{ or } -45°$$

I went ahead and gave a second solution as well.

Now, part (c):

$$\frac{1}{2}\sin x = 4$$
$$\sin x = 8$$
$$x = \sin^{-1}(8)$$

What happened? If you used your calculator you should have gotten an error message. There is no angle whose sine is 8. Remember that the range of the sine function is $-1 \le y \le 1$. Eight is not in that range. Therefore, this equation has no solution.

In the last few examples I have been using an exponent of -1 to represent the inverse of the trig functions. This is standard notation but sometimes people mistakenly think it is an exponent. Remember: $\sin^{-1}x \neq \frac{1}{\sin x}$. The first one, $\sin^{-1} x$ is the inverse. The second, $\frac{1}{\sin x}$, is the reciprocal. Each of the three trig functions you know, sine, cosine and tangent, have a reciprocal function, cosecant, secant and cotangent, but you don't need to worry about these. These functions were much more useful before people had access to calculators.

Another topic from before calculators is **trig identities**. When trying to solve equations with trig functions or trying to find parts of triangles, sometimes the trig expressions became very messy. To help simplify things, mathematicians came up with a long list of trig identities. These were facts that had been proven true and could be used to rewrite one trig expression as another. For instance, $\sin 2x = 2\sin x \cos x$. Before calculators, students spent grueling hours proving these identities. These proofs would be similar to proofs you might have done in geometry, where each step had a reason and the steps would follow logically, but would also be somewhat algebraic. Substitution was used a lot.

With the widespread availability of calculators, students can verify an identity similar to the one above by graphing both sides of the equation to show that they are equal. Try it to be sure you get the same thing. In the **More to Try** problems I have included just one for you to verify. There really isn't much for you to do other than graph it on your calculator. If you want more practice, see an older trig book, circa 1975. You will find <u>pages</u> of trig identities to verify.

This unit is an easy one for you to review and practice on your own. Since most of the unit involves finding parts of triangles, you can quickly create your own problems. You can check them by reworking them or by drawing the picture to scale. Another very helpful practice is to write sample questions with your team and have everyone try solving them. By comparing your solutions you will have a great way to check and you will all be learning a great deal from the process.

UNIT 9
MORE TO TRY

For each problem below, parts of ΔABC are given. Find the parts asked for using whatever method you prefer.

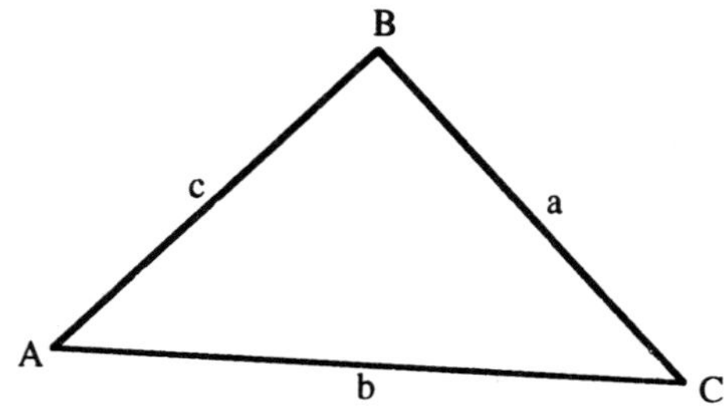

1. $m\angle A = 40°$, $m\angle B = 88°$, $a = 15$. Find b.
2. $m\angle B = 75°$, $a = 13$, $c = 14$. Find b.
3. $m\angle B = 50°$, $m\angle C = 60°$, $b = 9$. Find a.
4. $m\angle A = 62°$, $m\angle C = 28°$, $c = 24$. Find a.
5. $m\angle A = 51°$, $c = 8$, $b = 12$. Find a.
6. $m\angle B = 34°$, $a = 3$, $b = 4$. Find c.
7. $m\angle B = 96°$, $m\angle A = 32°$, $a = 6$. Find c.
8. $a = 15$, $b = 12$, $c = 14$. Find $m\angle C$.
9. $m\angle A = 30°$, $m\angle C = 60°$, $a = 8$. Find b.
10. $a = 8$, $b = 16$, $c = 7$. Find $m\angle C$.
11. $a = 9$, $b = 12$, $c = 15$. Find $m\angle B$.
12. $m\angle C = 18°$, $m\angle B = 54°$, $b = 18$. Find c.
13. $m\angle C = 76°$, $a = 39$, $b = 19$. Find c.
14. $a = 34$, $b = 38$, $c = 31$. Find $m\angle B$.
15. $m\angle C = 84°$, $m\angle B = 23°$, $c = 11$. Find b.

Find the area of each triangle.

16.

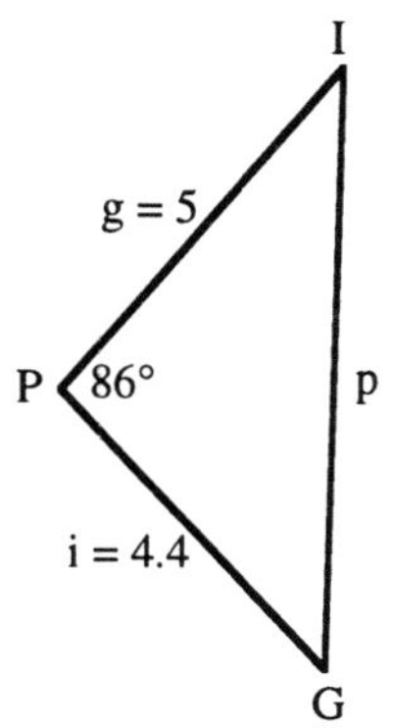

17.

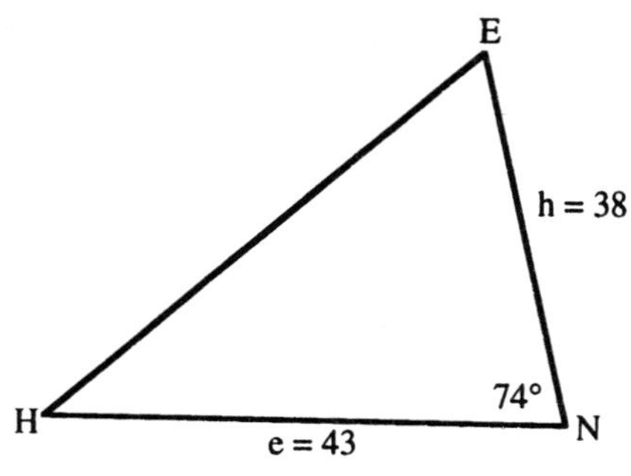

18.

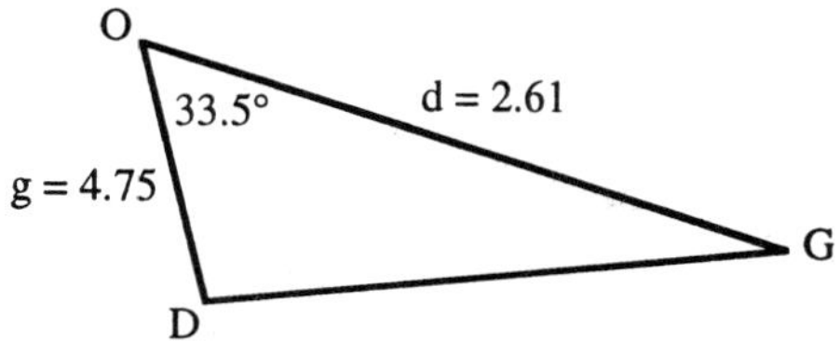

19.

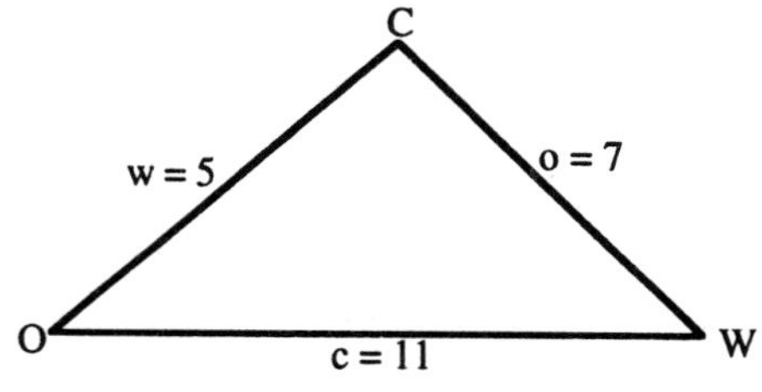

Solve the following equations for x. Assume x is in degrees.

20. $3 \sin x + 7 = 7$

21. $-4(2 + \cos x) = -10$

22. $(3 + 5 \sin x)(3 - 5 \sin x) = 0$

23. $\frac{.75 + \cos x}{3 \sin x} = 0$

24. Investigate $y = \frac{1}{\sin x}$.

25. Graph each of the following and based on the results, write a conjecture.

$y = \cos 2x$ $\qquad\qquad$ $y = 2\cos^2 x - 1$

SAT PREP

1. Six consecutive integers sum to 327. Find the largest of these integers.

2. If $x^3 - y^3 = 64$ and $x^2 + xy + y^2 = 4$, then what is the value of $x - y$?

3. On an exam, a class with m students earn an average (arithmetic mean) of 73. In another class, n students earn an average of 88. The combined average of all the students in the two classes is 85. What is the value of $\frac{m}{n}$?

4. Suppose x and y are different positive integers and that $x + y = 5$. What is the greatest possible value of $3x + 4y$?

5. The product of five and n is a multiple of 10. If the product is positive but less than 70, what is the largest possible value of n?

UNIT 10

Policy Making

PROBABILITY

UNIT 10
POLICY MAKING AND PROBABILITY

The odds are you have seen probability before this unit. But, you may not have encountered problems as complex as some of these. It will help to remember those simpler, similar probability problems you can already easily solve. For example: What is the probability of rolling a six? What is the probability of drawing a King of Hearts out of a deck of cards?

The problems in this unit go much further than earlier problems. Two things you will know well by the end of this unit are tree diagrams and area diagrams. These diagrams help to clarify the situation described in the problems, and sometimes, once you've drawn the diagram you can quickly read the answer from it.

Let's begin with a couple of simple examples to make sure everyone knows how the diagrams are used. At times I will present a slightly different approach to the problem than your text. If my explanation doesn't help you, stick with the one in the book.

1. Tess has three shirts, a red, a blue, and a green, and two pairs of pants, one black and one navy. If she chooses a shirt and a pair of pants at random, what is the probability that she is wearing

a) a red shirt? **b) black pants?**

c) a green shirt with black pants?

Solution.
This problem might be very simple for some of you, and you may be able to solve it without the use of a diagram. However, let's make sure that you know how to set up a simple diagram before we do more complicated problems.

We use a diagram to help us see all the possibilities that can occur. All the possibilities is known as the **sample space**. Sometimes making a list can be too long and tedious. Other times there might be too much information to try to include all of it in a list. The tree diagram helps organize and display the information so that we can see all the possible outcomes.

In many cases, we really don't need to know <u>all</u> the possible outcomes, but we do need to know <u>how many</u> outcomes there are, that is, we need to know the number of items in the sample space. This particular problem doesn't ask us what all the different outfits are. But it does ask for the probability of specific outfits occurring. We need to know how many total possibilities there are to calculate the probabilities. We start by considering the first "choice." Tess can select one of three possible shirts. I represent this with a tree diagram with three branches, each branch representing one of the possible outcomes. Once she chooses a shirt, she must next choose a pair of pants. She has only two choices, but these two choices occur with <u>each</u> one of the possible shirts.

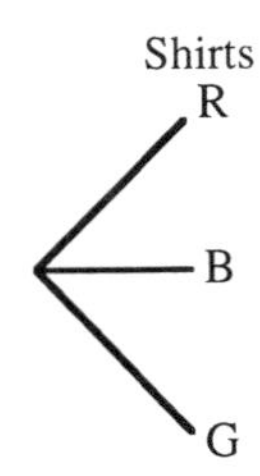

We represent this by showing additional branches coming from each of the original branches. Although we weren't asked for a list of outfits, you can see now that the list would be easy to give. She has a red shirt with black pants, red shirt with navy pants, blue shirt with black pants, etc. You could also quickly list outfits with abbreviations: RBl, RN, BBl, BN, GBl, GN. We can also quickly see that there are six possibilities. Now let us answer the questions posed. What is the probability of Tess wearing a red shirt? There are two outfits out of the six which have a red shirt, so the probability is $\frac{2}{6} = \frac{1}{3}$. The probability of Tess wearing black pants is $\frac{3}{6} = \frac{1}{2}$ because half of her outfits have black pants. Lastly, the probability of a green shirt with black pants is $\frac{1}{6}$ because there is only one outfit out of the six like that.

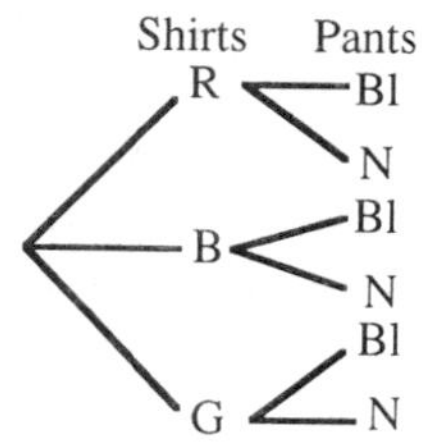

Tree diagrams work nicely to help organize information, but only when there isn't that much of it. If Tess were to have eight shirts and twelve pairs of pants, imagine drawing <u>that</u> tree diagram! The diagram would become one big cluttered mess and would be hard to read. That's when an area diagram can be especially helpful. Let's start with the most common area diagram: one representing the roll of two dice.

2. A red die and a green die are rolled together. What is the probability that the sum of the two dice is

a) six? **b) even?** **c) greater than 4?**

Solution.

The dice really don't have to be two different colors, as you might already know, but it does help keep track of the different outcomes. Each die has six different possible outcomes, and each is equally likely. (That means you are just as likely to roll a 1 as any other number on the die.) Our diagram lists all possibilities for the red die along one side and all the possibilities for the green die along an adjacent side. Notice how each little square is the same size. This represents the fact that each outcome is equally likely. Each square represents a different outcome, so the square labeled A, for instance, represents rolling a three on the red die and a five on the green die. There are 36 little squares and there are 36 possible outcomes when these two dice are rolled.

Red \ Green	1	2	3	4	5	6
1						
2						
3					A	
4						
5						
6						

Since the problem asks us about the sum of the two dice, we can fill in each square with the sum. This will help us answer the questions.

Red \ Green	1	2	3	4	5	6
1	2	3	4	5	6	7
2	3	4	5	6	7	8
3	4	5	6	7	8	9
4	5	6	7	8	9	10
5	6	7	8	9	10	11
6	7	8	9	10	11	12

Part (a): out of the 36 possibilities, how many times is the sum 6? The diagram shows five times, so the probability is $\frac{5}{36}$. Part (b): how often is the sum even? Eighteen, half of the time, so the probability is $\frac{1}{2}$. Part (c): how often is the sum greater than 4? Don't include equaling four! Thirty times for a probability of $\frac{30}{36} = \frac{5}{6}$.

That's basically how these diagrams work. You might be wondering why the second diagram is referred to as an area diagram. You may not have realized that you were working with area in the dice problem. But if the grid above represented a dart board and we asked the probability of throwing a dart and hitting an even number, the probability would depend on how much area the even numbers take up compared to the total area of the dart board. Because the outcomes are equally likely in the dice problem, and since each small square is a 1 by 1 unit square, counting squares and counting area are the same thing.

Now let's look at a more complicated problem, PM-43. We will use an area diagram to solve this problem as well as some other questions I will ask.

PM-43. A spinner comes up red 25% of the time and green 25% of the time. The rest of the time it lands on blue.

a) Draw an area diagram for spinning twice, and shade the region on your area diagram corresponding to getting the same color twice. A neat and accurate diagram on this problem will help in understanding the next several problems. What are the dimensions of the whole diagram?

b) What is the probability that both spins give the same color?

c) If you know that you got the same color twice, what is the probability the color was blue?

Solution.

Drawing a diagram is somewhat like converting this problem of spinning a spinner twice to throwing darts at a dart board. When spinning the spinner twice we have several different possible outcomes: red-red, red-green, red-blue, green-red, green-green, green-blue, blue-red, blue-green and blue-blue. These outcomes are not equally likely so we cannot just create a dart board with nine congruent squares on it. The size of each square must reflect the probability of getting that particular color combination.

When we made the grid for the dice problem, we listed all the possibilities along the sides of the grid, one side representing the red die and an adjacent side representing the green die. Similarly in this area diagram, one side will represent the first spin and the adjacent side will represent the second spin. The big difference is we cannot divide the sides up into equal-sized pieces. In this problem the chances of getting blue on a spin are different from getting a red or green. But, we do know what the chances are. Half of the time, blue should come up, so half of the side should be blue. Red and green each come up 25% of the time so one-fourth of the side should be red and another one-fourth green. We label the adjacent side for the second spin the same way. Doing this splits the "dart board" into nine squares but not nine congruent squares.

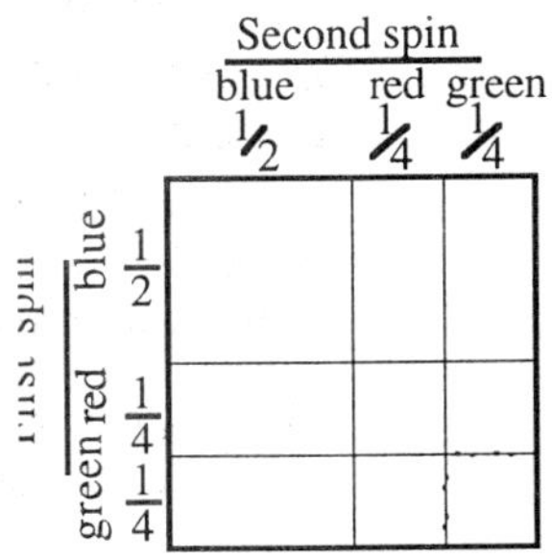

As you might notice, if you throw a dart at this dart board you are most like to hit the region labeled blue-blue, and that was most likely to come up in two spins of the spinner. What is the probability of hitting this square? We can find the fraction of the total area that is represented by the blue-blue square, and that fraction will be the probability. The sides of the blue-blue square are each $\frac{1}{2}$ units long, so the area is $\left(\frac{1}{2}\right)\left(\frac{1}{2}\right) = \frac{1}{4}$. The large square is a 1 by 1 square, with area 1 square unit. Therefore the probability of getting blue-blue is $\frac{1}{4}$.

What is the probability that the two spins will be the same color? That's the same question as: what is the probability of hitting blue-blue, red-red, or green-green? These three areas are shaded slightly on the dart board. To find the probability we must find the area of the total shaded region.

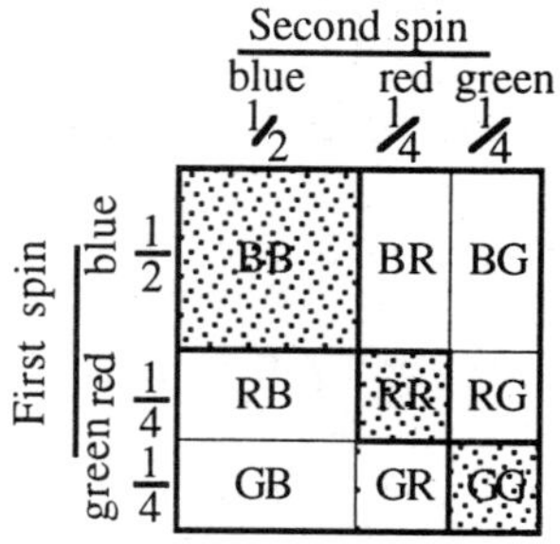

$$\begin{aligned}\text{Total Area} &= A(BB) + A(RR) + A(GG)\\ &= \frac{1}{4} + \left(\frac{1}{4}\right)\left(\frac{1}{4}\right) + \left(\frac{1}{4}\right)\left(\frac{1}{4}\right)\\ &= \frac{1}{4} + \frac{1}{16} + \frac{1}{16}\\ &= \frac{4}{16} + \frac{1}{16} + \frac{1}{16} = \frac{6}{16} = \frac{3}{8}\end{aligned}$$

Note: we actually take the area of the shaded region <u>divided</u> by the total area, but since the total area is 1, dividing by 1 yields the same results.

Part (c) asks a slightly different question: If you know you got the same color on both spins, then what is the probability that they were both blue? This is an example of **conditional probability**. An extra condition has been added to the question. To answer this, we look at a smaller sample space. We already know that the two colors are the same. That's similar to saying our dart has landed somewhere on this new board, shown at right. The total area of this board is $\frac{1}{4} + \frac{1}{16} + \frac{1}{16} = \frac{3}{8}$ as we saw above. How much of this area is blue-blue? The area of blue-blue is $\frac{1}{4}$ so the probability that we got blue-blue is

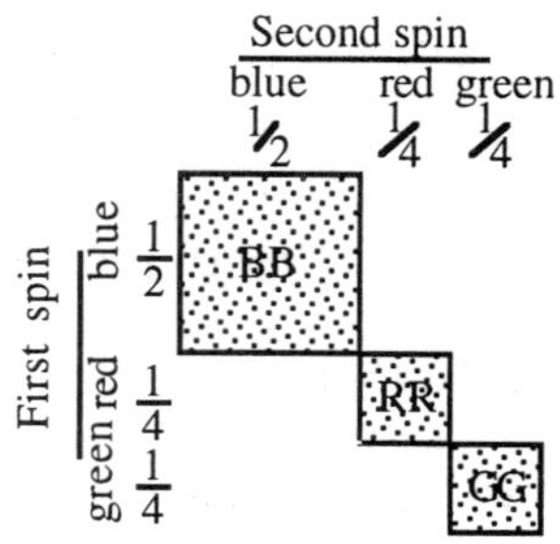

$$\frac{\frac{1}{4}}{\frac{3}{8}} = \frac{1}{4} \div \frac{3}{8} = \frac{1}{4} \cdot \frac{8}{3} = \frac{2}{3}$$

Here is another conditional probability question for you to try using this same diagram. Suppose we know that we got at least one green. What is the probability that we actually got two greens? Now we need to look at the part of the board that has at least one green. I shaded that part at right. This gives us a new dart board to consider, shown next to the first. The total area of the new dart board is

$2\left(\frac{1}{4}\right)\left(\frac{1}{2}\right) + 3\left(\frac{1}{4}\right)\left(\frac{1}{4}\right) = \frac{1}{4} + \frac{3}{16} = \frac{7}{16}.$

How much of this new board actually has green-green? $\frac{1}{16}$. So now we can calculate the probability.

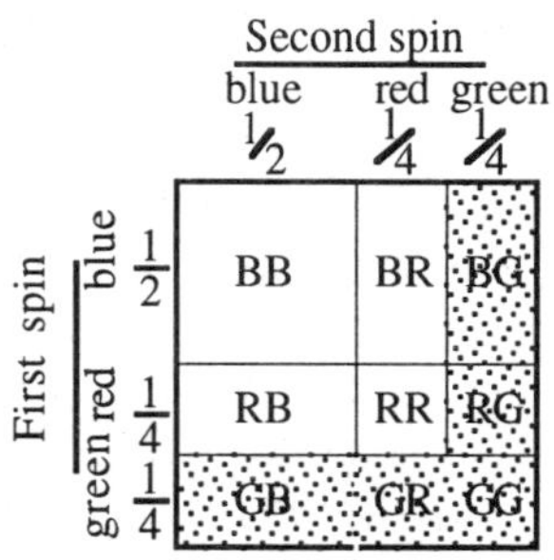

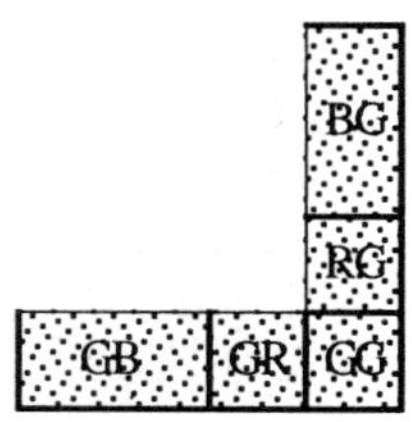

$$\frac{\frac{1}{16}}{\frac{7}{16}} = \frac{1}{16} \div \frac{7}{16} = \frac{1}{16} \cdot \frac{16}{7} = \frac{1}{7}$$

If you find that you are understanding the probability questions but are having trouble with the fractions, here's a little trick that might help with some of the calculating. When you are beginning to draw your area diagram, draw it on graph paper. Start with a square (or rectangle, as we will see a little later) and choose clever dimensions. For instance, in the last problem, the probabilities that were given for blue, red and green where $\frac{1}{2}$, $\frac{1}{4}$, and $\frac{1}{4}$. The common denominator for these three fractions is four. I could have started with a four by four unit square as my dart board. Then I can easily choose the appropriately sized areas for the blue, red and green possibilities. Next, label each square correctly (note that there are four squares labeled BB). Since all the squares are the same size now, we can calculate the probabilities by counting the number of squares just as we did with the red and green dice problem. You can try this method on some of the other problems. On problems such as PM-45, though, this won't work very well since the probabilities are unknown to us. We will look at that situation next.

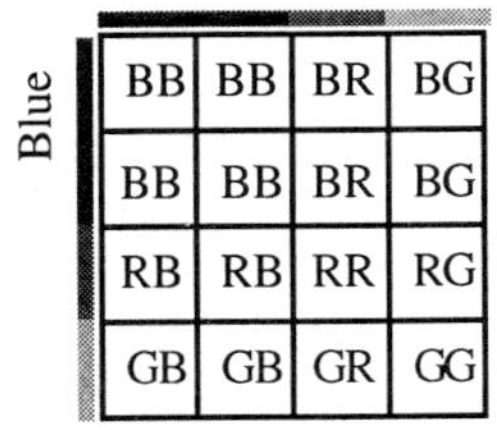

PM-45. A spinner has just two colors, red and blue. The probability the spinner will land on blue is x.

a) **What is the probability it will land on red?**

b) **Sketch an area diagram for spinning twice.**

c) **When it is spun twice, what is the probability it will land on the same color both times?**

d) **Given that it lands on the same color twice, what is the probability that it landed on blue both times?**

Solution.

This problem may look like different since the probability of the spinner landing on blue is **x**. But we can do this problem by remembering the simpler, similar problems we just did. There are only two colors on the spinner: blue and red. If the probability of landing on blue were one-half, what would be the probability of landing on red? One-half as well. If the probability of landing on blue were one-fourth, then what would be the probability of landing on red? Three-fourths. What if the probability of landing on blue were three-eighths, what would be the probability of landing on red? Five-eighths. Are you seeing a pattern? So if the probability of landing on blue were **x**, what would be the probability of landing on red? **1 - x**. Try some more numbers if you are not convinced of this. We can start our area diagram (our dart board) but since we are using variables, the diagram is really only a sketch. We cannot be too specific, but we should not draw the small squares all the same size. We don't know if the events are equally likely. Now we can answer the questions. What is the probability of getting the same color twice (i.e. of hitting BB or RR)? This dart board is a 1 by 1 unit square so it has area 1. What fraction of it is made up by BB and RR? That area is

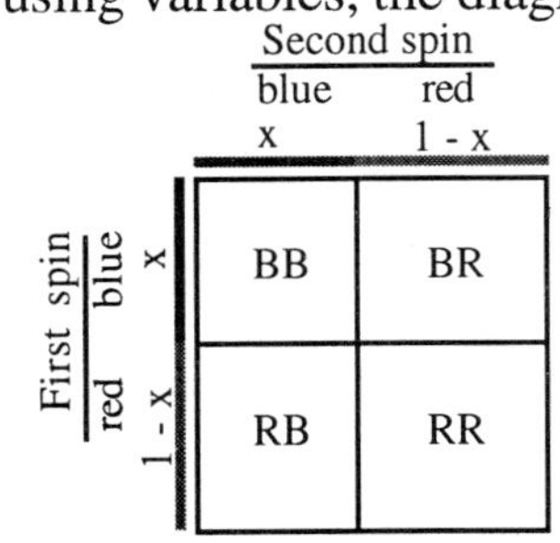

$$\begin{aligned} A &= x(x) + (1 - x)(1 - x) \\ &= x^2 + 1 - 2x + x^2 \\ &= 2x^2 - 2x + 1 \end{aligned}$$

Don't let this strange looking probability alarm you. We did this problem just as we did the others so we must trust our mathematical abilities.

If we know we got the same color twice, what is the probability that we actually got two blues? Now we look at our new dart board, made up of the same color outcomes. The new dart board has a total area of $2x^2 - 2x + 1$ (we just calculated it above, remember?) and the part of it that is BB has area x^2. Therefore the probability is $\frac{x^2}{2x^2 - 2x + 1}$. This is the same process as the preceding problems, except we are using variables.

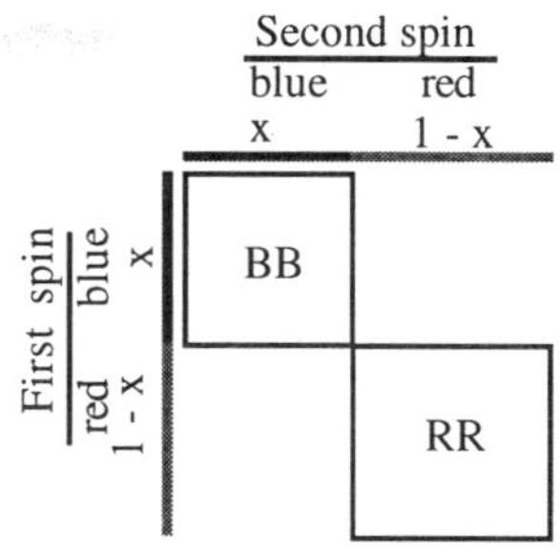

Let's see what happens when we perform different actions depending on the outcomes. You saw similar problems in PM-56 and PM-57.

3. Deniz is trying to create a fair game for the Math Club's booth at the spring carnival. The player first rolls a die. If the outcome is less than three, the player chooses a ping pong ball from a box. There are ten balls, numbered one through ten. If the number on the ping pong ball is also less than three, the player wins a prize. If the player's die has a number greater than or equal to three, the player chooses a ping pong ball from a different box containing only four balls numbered one through four. If that number is also greater than or equal to three, the player wins a different level of prize. Of course, in order to be sure the game is fair she needs to know the probabilities of each of the outcomes. Draw an area diagram which shows all the outcomes with their probabilities.

Solution.

A lot is going on here. Let's take it one step at a time and we will see that this problem is similar to the earlier ones, and that we can do this one too. To begin with let's discuss the roll of the die. If we get a number less than three (< 3) we will do something different than if we get a number greater than or equal to three (≥ 3). There are two numbers less than three: one and two, out of six possible outcomes. Therefore the probability of getting a number less than three is $\frac{2}{6} = \frac{1}{3}$. We show this on our diagram, splitting the side as one-third and two-thirds. Now what happens next depends on whether we are in the top part of the dart board or the bottom. If we got a number less than three (and are in the top part) then we choose a ping pong ball from a box of ten of them. That means we have ten equally likely possibilities. Across the top of the diagram, then, I will divide the side up into ten equal sections. Note: because of space, I'm not writing the probability of each number. You will just have to remember that each has a probability of $\frac{1}{10}$ of occurring. So what happens if we didn't get a number less than three (i.e. we are in the bottom part of our diagram)? Then we choose from a box of only four balls, numbered one through four. Each of

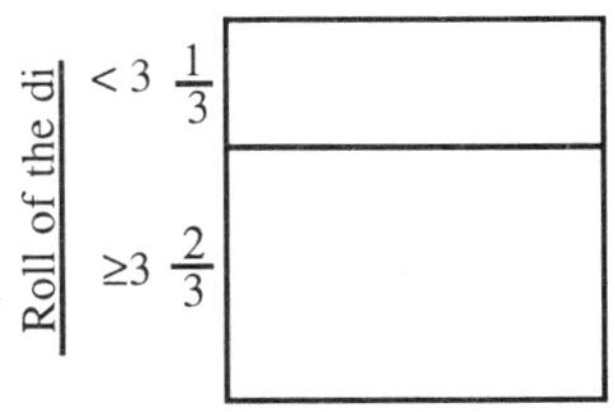

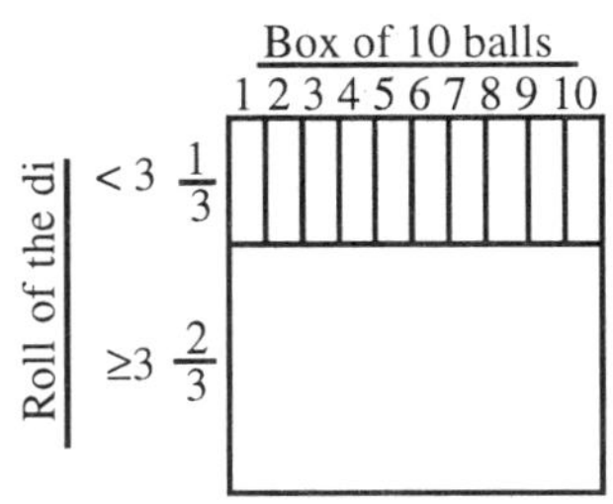

these balls is also equally likely. I will represent this along the bottom of the diagram. Now the question is, what are the probabilities of each of these outcomes occurring? The area of each skinny rectangle across the top is $\left(\frac{1}{3}\right)\left(\frac{1}{10}\right) = \frac{1}{30}$. The area of each of the four rectangles across the bottom is $\left(\frac{2}{3}\right)\left(\frac{1}{4}\right) = \frac{1}{6}$. Knowing this, let's find the probability of winning a prize. The player wins a prize two ways. One way is if the player gets numbers less than three in both instances, with the die and with a ping pong ball. This means the player would have to get a one or a two on the ping pong ball. The second way is if he gets numbers greater than or equal to three in both instances. This means he must get a three or a four on a ping pong ball. These two winning possibilities are indicated in the figure at right in the shaded areas of the diagram. To find the probability of winning, then, is the same as pretending this is a dart board and finding the probability of hitting the shaded region. The area of the shaded region is

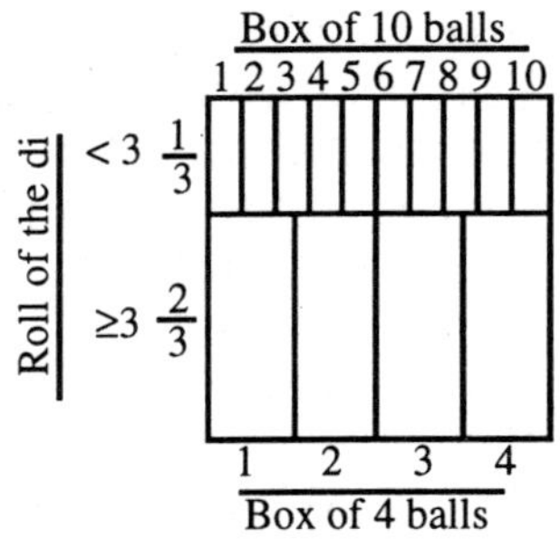

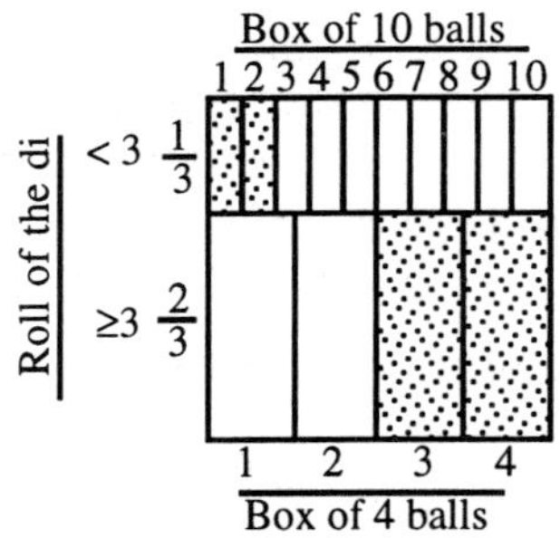

$$A = 2\left(\frac{1}{3}\right)\left(\frac{1}{10}\right) + 2\left(\frac{2}{3}\right)\left(\frac{1}{4}\right)$$

$$= \frac{1}{15} + \frac{1}{3} = \frac{6}{15} = \frac{2}{5}$$

Once you feel comfortable with area diagrams, you should be able to tackle just about any probability problem. You might have friends who have heard rules about adding probabilities and multiplying probabilities. You might ask your friends if they always know when to add and when to multiply. That is often what is confusing about these problems. Area diagrams help you see what to do and why the answer is what it is. You may want to show your friends how to do these diagrams.

The last thing we will look at about probability in this unit is **expected value**. Many people confuse expected value with the term "most likely." They really are not the same thing. Try the following problem to see if you understand these terms.

4. In a simple coin game, the player flips a coin twice. If both outcomes are heads, the player gets one point. If both outcomes are tails the player gets two points. If the two flips have different outcomes (i.e. heads-tails or tails-heads), the player gets zero points. What is the most likely outcome? What is the expected number of points a player is going to get?

Solution.

The first thing we need to do is make sure we know all the outcomes and what the probability of each outcome is. To do this, I will use an area diagram, although a tree diagram would work just as well here. Each flip has two equally likely outcomes. Our square is divided into four congruent smaller squares. The two shaded squares represent the outcomes in which the player will get zero points. The diagram makes it clear that half of the board produces zero points. One-fourth of the time a player will get heads-heads for one point and one-fourth of the time a player will get tails-tails for two points. Therefore, the most likely outcome is for a player to get zero points.

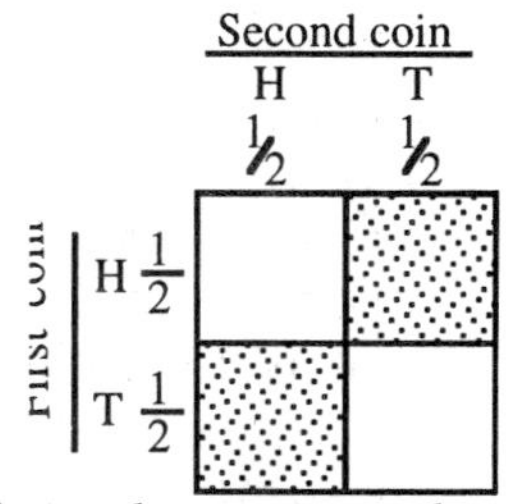

But that's not the expected value for this game! Let me explain further what expected value is. Suppose you played this game hundreds of times and you kept totaling your points. When you decided to end your playing, you might have a score in the two or three hundreds. Then, as you look back, what was the average number of points you got each time you played? That's what expected value is. So let's suppose you play this game four hundred times, keeping track of your points. You would probably get zero points half the time, so there would be 200 zeroes in your list. About one-fourth the time you would get one point, so there would be about 100 ones in your list. And there would also be about 100 twos in your list. What is the total? Three hundred points. Since you played four hundred games you averaged only $\frac{300}{400} = \frac{3}{4}$ points each time you flipped the coins. The expected value for the number of points for this game is $\frac{3}{4}$. Notice that you get the same result if you examine playing the game four times. If each region came up once, you would have 3 points in four flips, which is also $\frac{3}{4}$.

Note: if you are having trouble with PM-68, the sector of the circle worth \$100 is one-eighth of the circle. Use that for its probability. Also, if you can't seem to answer part (c) asking for the expected value, try doing what I just did. Imagine playing the game several hundred times.

This last topic I will consider is **Markov Chains**. Although this name doesn't imply it, this is merely using matrices and matrix multiplication to model probability applications. This method is clearly explained in PM-99 through PM-104. Here is a similar problem.

5. At Bueller High School, the students LOVE school. Some of them, however, have a problem with tardiness. If a student is late to a class on one day, the probability that he will be late the next day is 0.2. If he is on time on one day, the probability that he will be late the next day is 0.5. On the first day of school, 175 of the 1000 students are late.

a) What will happen on the second day of school?

b) What about the third day of school?

c) What will happen by the end of the school year, 180 days later?

Solution.

As we learned earlier, matrices are just tables to organize and store information. First we will make a matrix to tell us how many students were late and how many were on time. We can do this with a 1 x 2 matrix, let's call it A.

$$A = [175 \quad 825] \, .$$

We also write another matrix, sometimes called a transition matrix, to hold the probabilities. To fill in the numbers correctly on this transition matrix, T, I will first create a table with descriptive headings.

	Late the next day	**On time the next day**
Late the first day	**0.2**	**0.8**
On time the first day	**0.5**	**0.5**

By looking at the first row, this table tells us that if a student is late the first day, the probability that he will be late the second day is 0.2. If a student is late on the first day, the probability that he will be on time on the second day is 0.8. Notice that the probabilities in the rows sum to 1. There are only two possibilities here: either a student is late or on time. If the probability that a student is late is 0.2 the probability of the other option (on time) is 1 - 0.2 = 0.8. This table becomes this matrix:

$$T = \begin{bmatrix} 0.2 & 0.8 \\ 0.5 & 0.5 \end{bmatrix} .$$

To determine what will happen on the second day of school, we multiply AT.

$$AT = [175 \quad 825] \cdot \begin{bmatrix} 0.2 & 0.8 \\ 0.5 & 0.5 \end{bmatrix} = [447.5 \quad 552.5]$$

Thus, on the second day of school, approximately 448 students are late and the rest (552) are on time. To find out what will happen on the third day, we multiply by T again.

$$ATT = AT^2 = [447.5 \quad 552.5] \cdot \begin{bmatrix} 0.2 & 0.8 \\ 0.5 & 0.5 \end{bmatrix} = [365.75 \quad 634.25]$$

By the third day, about 366 students are late and 634 are on time.

As you can see, the numbers are bouncing around a little. What do you think will happen by the end of the school year? You might think that the numbers will keep bouncing around, never really settling down anywhere, but if you take the answer each time and then multiply by T again, you will see the numbers leveling off. In fact, you can find the answer directly by computing AT^{180}. This is a good chance to use your calculator to see if you can perform matrix multiplication properly.

$$AT^{180} \approx [384.6 \quad 615.4]$$

By the end of the school year, approximately 385 students are late and 615 are on time.

Unit 10 has many problems. Hopefully you will see the connections between the problems and that will aid in your learning and understanding. By understanding the area diagrams and tree diagrams you will be making the connections you need. The next unit will look at probability from a different perspective.

UNIT 10
MORE TO TRY

Complete the square to find the vertex and sketch the graph of the following.

1. $y = x^2 - 8x + 25$
2. $y = x^2 + 8x + 10$
3. $y = 3x^2 - 6x + 15$
4. $y = -4x^2 - 24x - 38$
5. $y = \frac{1}{2}x^2 + 6x + 25$
6. $y = .1x^2 + 1.4x + 9.9$
7. $y = 3x^2 - 54x + 237$
8. $y = -4x^2 + 20x - \frac{43}{2}$

When the chef at Honeypenney's House of Pasta prepares side dishes, she doesn't put much thought into it. She randomly chooses one of three vegetables, either broccoli, carrots or peas. Then she randomly chooses a type of bread stick, either garlic or rosemary.

9. Use a tree diagram to show all the possible pairs of vegetable with bread.

10. Using your tree diagram, what is the probability of getting broccoli with garlic breadsticks?

11. What is the probability of getting carrots and rosemary bread sticks?
12. Suppose you really want garlic breadsticks so you slip the waiter a five to be sure you get them. If you are certain you will get garlic breadsticks, what is the probability your order will be peas with garlic breadsticks?

A red die and a green die are rolled. What is the probability of

13. the sum being greater than 7?

14. the sum being a multiple of 3?

15. the sum being less than 12?

A spinner has 40% of it colored blue. Red and green make up the rest of the spinner with red and green taking up the same amount of space.

16. What is the probability of the spinner landing on red?

17. Draw an area diagram representing spinning the spinner twice. What is the probability of getting the same color on both spins?

18. Suppose you know that the spinner landed on the same color twice. What is the probability that the color was red?

19. Suppose you know that at least one time the spinner landed on blue when it was spun twice. What is the probability that the other color was also blue?

20. In a dice game , the player rolls a die. If he gets a one or two, he stops. If he gets any other number, the player gets a point and rolls one more time. If he gets a one or two on the second roll he adds no more points to his score; any other number and he adds one more point and his turn is over. What is the probability of getting

a) zero points? b) one point? c) two points?

21. What is the expected value of the game just described in the last problem?

The Health Department is developing a two-stage procedure to relieve the symptoms of a rare blood disorder. The first procedure has a 70% probability of causing some relief (which is considered a success). If it does succeed the second procedure has a 90% probability of success. If the first procedure doesn't succeed, the second procedure has only a 40% probability of success. What is the probability that

22. both procedures are successful?

23. both procedures fail?

24. the first fails and the second succeeds?

25. If you know the second procedure was successful, what is the probability that the first procedure was successful?

In greater Ferris, the city council is very particular about what the citizens of Ferris can and cannot do. For instance, the city only allows two types of toothpaste to be sold in their stores: Coldstate and Crust. Each time a Ferrisite purchases a tube of toothpaste, she has a $\frac{1}{3}$ chance of choosing the same brand on her next purchase. Of the 8,000 people living in Ferris, 3028 bought Crust the last time they bought toothpaste.

26. How many will by crust on their next visit?

27. How many will buy Coldstate on their fifth visit?

28. What will happen in the long run?

SAT PREP

I've included only five sample problems here, with five more in the next section.

1. In the figure at right, line *l*, *m*, and *n* are parallel. What is the value of x + y?

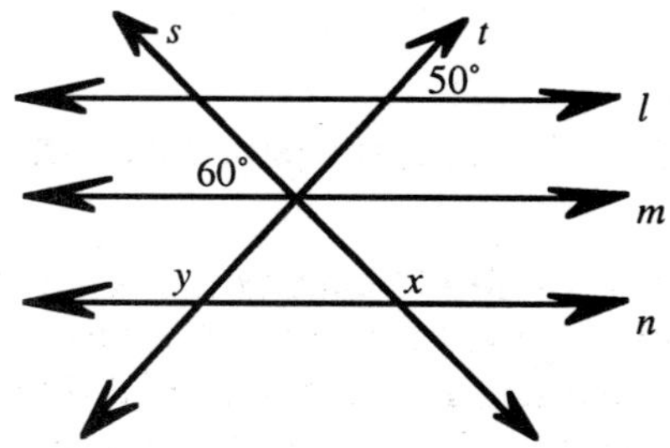

a) 250 b) 230 c) 180

d) 110 e) 70

2. A repeating decimal has a block of numbers that repeat indefinitely. In this repeating decimal

$$.\overline{13579} = .1357913579\ldots,$$

which digit is in the 3,000th place to the right of the decimal point?

a) 1 b) 3 c) 5 d) 7 e) 9

3. In the figure at right, A and B are the centers of the two circles. If each circle has an area of 5, what is the area of the rectangle?

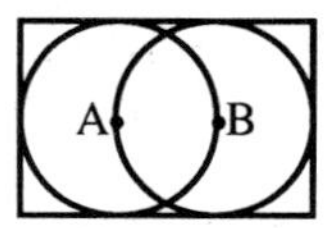

a) 10 b) $10 - \frac{5}{\pi}$ c) $\frac{50}{\pi}$ d) $\frac{30}{\pi}$ e) $\frac{25}{\pi}$

4. There are four roads from A town to B town and three roads from B town to C town. If Gilbert travels from A town to C town and back, passing through B town in both directions, and <u>never</u> passing over the same road twice, how many different routes for this trip are possible?

a) 12 b) 18 c) 24 d) 36 e) 72

5. If a • b • c • d = 0 and b • c • d • e = 1, which of the following must be equal to zero?

a) a b) b c) c d) d e) e

UNIT 11

The Ice Cream Shoppe

COUNTING AND SAMPLING

UNIT 11
COUNTING AND SAMPLING

You will notice quickly how this unit is different from the last. Although both units are about probability, each takes a different approach. Unit 10 emphasizes the use of diagrams (tree and area) to solve probability problems, many of which were conditional probability questions. We often listed all the possible outcomes. In this unit, the emphasis is on counting how many items are in the **sample space**, the total collection of possible outcomes. We count how many items are in the sample space and how many items are in the particular event we are interested in. With these two numbers we can determine the probability if we want to. Many times we are just interested in the number of occurrences.

We will develop several ways to do this counting and will define some new mathematics along the way. Most of the problems in this unit will have the same "sound," asking questions beginning "How many ways . . ." Let's start with such a problem.

---

1. Mrs. Ferret's four children are about to line up for a family portrait. How many ways can the children line up?

Solution.
This problem can be done with a tree diagram by showing the possible order that the children can arrange themselves. For instance, if the children are labeled A, B, C, and D, then anyone of them could go first. If A were to go first, then either B, C, or D could go next. We could continue the branching for each one to see all the possible line ups. But, you might realize that we don't really care what those line ups look like or who is where in the line up. All we really care about is <u>how many</u> line ups there are. Certainly if we draw a tree diagram we can count how many there are, but that does seem like a lot of extra, unneeded work. Instead, let's use the idea of the tree diagram to help us count how many arrangements there are. Imagine that there are four positions, or seats, that each child will sit it. Someone must sit in the first seat. If you had to choose who would sit in the first seat, how many choices do you have? Four. Once that first child has sat down, how many choices are there for the second seat? Only three now. The next seat has only two choices and the last has only one possibility. This is what we were doing with the tree diagram, each time the number of branches decreased by one.

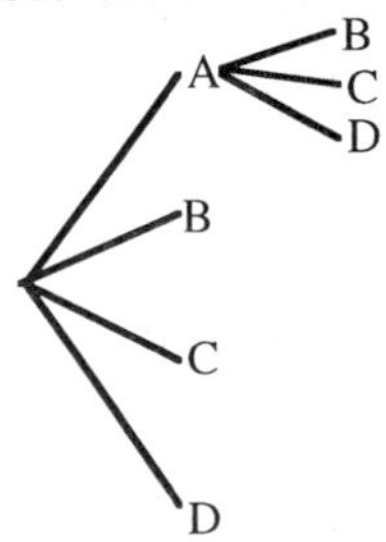

4	3	2	1
First seat	Second seat	Third seat	Fourth seat

The difference here is we are just counting the branches, not listing all the possibilities as well. We list the number of choices in a **decision chart** as shown above. The question is: what do we do with the numbers? (Look back at Unit 10 and read PM-2.) In the tree diagram above, each of the four original branches will have three branches coming off of it. That is a total of $4(3) = 12$ branches. Each of those paths will have two more branches coming off of it for a total of $12(2) = 24$. This gives us the answer. We multiply these numbers to find the number of arrangements: $4(3)(2)(1) = 24$. There are 24 possible portrait arrangements.

In the last example it is very important to notice that the order of the children matters. The family portrait would look different if the children were in a different order. This fact will become more important as we go through the unit. Let's look at another similar problem.

2. Twenty-three people have entered the pie eating contest at the county fair. The first place pie-eater (the person eating the most pies in fifteen minutes) wins a pie each week for a year. Second place will receive new baking ware to make his/her own pies, and third place will receive the *Sky High Pies* recipe book. How many different possible top finishes are there?

Solution.

Since the prizes are different for first, second and third place, the order of the top finishers matters. We can use a decision chart as we did in the last example to determine the number of different ways we can have winners. Although the winners aren't really "chosen" we can use the same idea. How many different people can come in first? Twenty-three. Once first place is "chosen" (i.e. removed from the list of contenders) how many people are left to take second place? Twenty-two. This leaves twenty-one possible third place finishers.

23	22	21
First	Second	Third

Just as with the branches on the tree diagram, we multiply these numbers to determine the number of arrangements: 23(22)(21) = 10,626. How would you like to draw <u>that</u> tree diagram!

Besides their phrasing, the last two problems have something else in common. We get the answer by multiplying several consecutive numbers. This occurs so often in counting problems that mathematicians defined an operation to simplify this. The operation is called **factorial** and the symbol is **!**. For example, when you see 4!, you read it as "four factorial" (not **"FOUR!"**) and it means to multiply the string of numbers beginning with 4 and decreasing down to 1: $4 \cdot 3 \cdot 2 \cdot 1$. So 4! = 24. Similarly, 20! is equal to $20 \cdot 19 \cdot 18 \cdot \ldots \cdot 3 \cdot 2 \cdot 1$, which if you use your calculator is a very large number, approximately $2.43 \cdot 10^{18}$. Note: you may have to read your calculator's manual to find out how to compute factorials. It is probably within one of the menu items. Also, you should see how large of a factorial your calculator can handle. My calculator can do 69!, responding in scientific notation of course, but gives an error for 70!.

Let's practice computing several factorials. Some you will be able to do without your calculator.

3. Compute the following.

a) $7!$ b) $\frac{20!}{18!}$

c) $\frac{100!}{98!}$ d) $\frac{75!}{70! \cdot 5!}$

Solution.

Part (a) is simple enough to be done on even a four-function calculator. $7! = 7 \cdot 6 \cdot 5 \cdot 4 \cdot 3 \cdot 2 \cdot 1 = 5040$. For part (b) you may want to use your factorial key if you haven't already, or you might see another way to do the problem. $\frac{20!}{18!} = 380$. For part (c) you might be thinking "What is she doing? She just said some calculators can't do factorials that big!" But there is a way to handle these numbers. If we think about what this fraction represents, we can simplify it.

$$\frac{100!}{98!} = \frac{100 \cdot 99 \cdot 98 \cdot 97 \cdot \ldots \cdot 3 \cdot 2 \cdot 1}{98 \cdot 97 \cdot 96 \cdot \ldots \cdot 3 \cdot 2 \cdot 1} = \frac{100 \cdot 99 \cdot \cancel{98} \cdot \cancel{97} \cdot \ldots \cdot \cancel{3} \cdot \cancel{2} \cdot \cancel{1}}{\cancel{98} \cdot \cancel{97} \cdot \ldots \cdot \cancel{3} \cdot \cancel{2} \cdot \cancel{1}} = 100 \cdot 99 = 9900$$

Try using this simplification whenever possible. It cuts down on the amount of work you must do. Part (d) can be done this way as well.

$$\frac{75!}{70! \cdot 5!} = \frac{75 \cdot 74 \cdot 73 \cdot 72 \cdot 71 \cdot 70 \ldots \cdot 3 \cdot 2 \cdot 1}{70 \cdot 69 \cdot 68 \cdot \ldots \cdot 3 \cdot 2 \cdot 1 \cdot 5 \cdot 4 \cdot 3 \cdot 2 \cdot 1} = \frac{75 \cdot 74 \cdot 73 \cdot 72 \cdot 71}{5 \cdot 4 \cdot 3 \cdot 2 \cdot 1} = 17{,}259{,}390$$

You might ask why anyone would want to use such incredible looking fractions. Well, when counting, they come up a lot in certain situations. One example we did earlier could have been written as a fraction. Look back at number 2 earlier. We found the answer by multiplying $23 \cdot 22 \cdot 21$. This actually comes from the fraction $\frac{23!}{20!}$. The decision chart we made did give us the correct answer. Many problems you will see throughout the rest of the unit will require you to look at the problem a little differently. Example 2 is an example of a **permutation**, an arrangement of things where order matters. The symbol for a permutation is ${}_nP_r$, where n is the number of things we are considering and r is how many we are arranging. In example 2, n, the number of things we were considering, is 23 and r, the number of things we were arranging, is 3. The formula for a permutation is

$${}_nP_r = \frac{n!}{(n - r)!}$$

so for our example we have: $\quad {}_{23}P_3 = \frac{23!}{(23 - 3)!} = \frac{23!}{20!} = 23 \cdot 22 \cdot 21$

And we get the same answer. Use permutations in the next example.

4. Fifteen students are trying out for a photo-shoot. In how many ways can you arrange

a) eight of them? b) two of them? c) fifteen of them?

Solution.

We can use a decision chart on each of these, but I will use the permutation notation and formula. In part (a), we have 15 things (students) with 8 of them to be arranged. (Order matters because if the students change their order, the picture will look different.) The permutation is

$${}_{15}P_8 = \frac{15!}{(15 - 8)!} = \frac{15!}{7!} = 15 \cdot 14 \cdot 13 \cdot 12 \cdot 11 \cdot 10 \cdot 9 \cdot 8 = 259{,}459{,}200$$

In part (b) the solution becomes: $${}_{15}P_2 = \frac{15!}{(15-2)!} = \frac{15!}{13!} = 15 \cdot 14 = 210$$

Part (c) poses a new problem: $${}_{15}P_{15} = \frac{15!}{(15-15)!} = \frac{15!}{0!}$$

What is 0!? We have been multiplying the numbers down to **one** when we compute factorials. If 0! = 0, then this problem has no answer because we would be dividing by 0, which we cannot do. So what does this equal? If we had used a decision chart to see how many ways 15 students can be arranged in 15 places, we would have $15 \cdot 14 \cdot 13 \cdot \ldots \cdot 3 \cdot 2 \cdot 1$ or 15!. (Verify this yourself! Make the chart to be convinced!) If ${}_{15}P_{15}$ = 15!, then 0! must equal 1. In fact, it does. This is another case of mathematicians defining mathematics to fit their needs. 0! is defined to equal 1 so that the other mathematics we are doing (like the case we are looking at) make sense.

The next question to consider is: when does order **not** matter? Order doesn't matter when you might be choosing people to form a committee, where everyone on the committee has an equal role. (If the committee is to have officers, say president, vice-president, etc., then the order of the members selected would matter and that would be a permutation.) Another example would be if a bag contained some small toys and you were going to reach into the bag and grab a handful. If you wanted to know the number of ways to grab a handful of five toys from a bag of 20, then the order wouldn't matter. When order does not matter, we are counting a **combination**. Try the next example.

5. In the annual homecoming parade, three students get to ride on the lead float. Seven students are vying for this coveted position. How many ways can three students be chosen for this honor?

Solution.

All three students selected ride on the lead float and no one of the three has any special title or power. This is a combination then, and not a permutation. Suppose the students are labeled A, B, C, D, E, F, and G. This means selecting A, then B and then C is the same as selecting B, then C, and then A, or any other arrangements. In fact, all the arrangement of A, B, and C could be lumped together. This makes the number of combinations **much smaller** than the number of permutations. The symbol for a combination is ${}_{n}C_{r}$, where n is the total number of items under consideration and r is the number of items we will choose. It is often read "n choose r". In this example we have ${}_{7}C_{3}$, 7 choose 3. The formula for combinations is similar to permutations, but we must divide out the similar groups.

$${}_{n}C_{r} = \frac{n!}{(n-r)!\,r!}$$

So in this example we have: $${}_{7}C_{3} = \frac{7!}{(7-3)!\,3!} = \frac{7!}{4!\,3!} = \frac{7 \cdot 6 \cdot 5 \cdot 4 \cdot 3 \cdot 2 \cdot 1}{4 \cdot 3 \cdot 2 \cdot 1 \cdot 3 \cdot 2 \cdot 1} = 35$$

You can see that you can greatly reduce this fraction.

Your calculator should be able to compute both permutation and combinations, but you may have to look for the correct buttons. If you can't find the buttons, don't fret. These fractions usually reduce so much you might be able to compute the numerical answer more quickly simply by multiplying the few numbers remaining after cancellation. Also, check with your teacher. It might be acceptable to leave your answer as ${}_7C_3$ or ${}_7P_3$ whichever is appropriate for the question. In the **More to Try** section at the end of this unit I will mix up the permutations and combinations so you will have to determine which type the problem is and make the decision as to how to solve it.

We have these convenient formulas, but there are some problems which don't use either of them. Then it might help to go back to decision charts. Let's look at an example.

---

6. Lacey is starting a new business and she needs to choose a name for it. She wants the name of her business to have three words in it and she feels she needs to choose the first letter of each of the three words first. Once she chooses the letters, then she will come up with a name. (For instance, if she chooses the three letters to be D, G, and I she might then call the business Doing Good International.) How many different three letter selections are there?

Solution.
It is important to point out that it is okay to repeat letters here. She may choose A, A, and A as her three letters. So although order does matter for most selections (DGI would give a different name then GID), this isn't a permutation since order doesn't matter if three of the same letter is chosen. This is a problem in which we allow repetition. How do we deal with this? Go back to our decision charts! We have three decisions to make. The first decision is: what will be the first letter? There are no restrictions on which letter it can be, so there are 26 choices for the first letter. Once we choose the letter, there are still 26 choices for the second letter, and also 26 choices for the third letter. This means there are $26^3 = 17{,}576$ possible selections.

26	26	26
First	Second	Third

Let's try one more example. You figure out what type it is.

---

7. The digits 0, 1, 2, 3, ..., and 9 are written on small square tiles.

a) How many ways can these tiles be arranged?

b) How many ways can these tiles be arranged to create a number less than 6 million?

c) These tiles are thrown into a bag. You are going to reach in and grab three numbers to throw away. How many ways are there to do this?

d) How many five digit numbers can be made with the remaining numbers?

Solution.
Part (a) is a permutation. We have ten numbers (yes, ten! Count them if you think there are nine.) and we want to arrange all ten of them. There are $_{10}P_{10} = \frac{10!}{(10 - 10)!} = 10! = 3628800$ ways to do this. Note: this could have also been done with a decision chart: ten slots, ten choice for the first slot. Once placed, there are only nine choices for the next slot, then eight, etc.

Part (b) is a little more tricky. If the number we create with these ten digits must be less than six million, the very first digit must be either 0, 1, 2, 3, 4, or 5. That means there are six choices for the first slot. Once chosen though, there are no other restrictions on the remaining numbers, other than once we use a number it is gone

$$\underline{6}\ \ \underline{9}\ \ \underline{8}\ \ \underline{7}\ \ \underline{6}\ \ \underline{5}\ \ \underline{4}\ \ \underline{3}\ \ \underline{2}\ \ \underline{1}$$

from the list. This means there are nine choices for the second slot, eight choices for the third slot, etc.

Did you recognize part (c) as a combination? Since we are grabbing three numbers to throw away, we don't care about the order. From these ten numbers we will choose three to throw away. This means there are $_{10}C_3 = \frac{10!}{(10 - 3)!3!} = 120$ ways to do this.

Now that we are left with seven digits (we threw three of the ten away), how many five digit numbers can we make? This is a permutation again since order matters (or you can use a decision chart). Here we have $_7P_5 = \frac{7!}{(7 - 5)!} = 2520$ different five digit numbers.

You might have noticed that I haven't asked any probability questions yet. Even though I haven't asked them, you still can do them. The probability question is usually not the tough part of the problem. The tough part is counting the number of items in the sample space and in the event we are interested in. Sometimes (especially in real world situations) there is no way at all to actually count the number of elements in the sample space. Then what is usually done is a sampling. You did such a sampling in IC-103 and IC-123. Biologists often will use a similar technique to determine a species population. Consider the following.

8. A park ranger tags 120 brown bears in his park. Six months later he catches (and releases) 75 brown bears. Only 42 of the these bears had tags. What does this tell the ranger about the brown bear population in his park? Explain

Solution.
It may seem drastically different to you, but this is very similar to drawing marbles out of a bag and using your observations to predict how many are in the bag. Let's look at a hypothetical situation first to understand the process. Suppose there were 200 brown bears in the park. Then when the ranger tagged 120 of them, he was marking 60% of them. From that point on, if a bear is chosen at random, the probability that it would be tagged would be 60%. Also, if several were chosen at random, you would expect about 60% of those chosen to be tagged.

Now let's use this idea in reverse. The ranger caught 75 brown bears and only 42 of them had tags. That's $\frac{42}{75} = 56\%$. If 56% of his catch was tagged, we would expect 56% of the entire population to be tagged. That means 120 is roughly 56% of the entire population. Algebraically,

$$120 = .56x$$
$$x \approx 214.286$$

Since we can't have .286 of a bear, the population is roughly 214 bears.

Counting problems take practice to master. I have included several in the problems which follow, but if you can find more to do you can improve your skill at them. You may have a friend who uses a different Algebra 2 book. Look in the index for permutations and combinations and you should be able to find lots of problems to try.

UNIT 11
MORE TO TRY

Simplify the following expressions.

1. $10!$
2. $\frac{10!}{3!}$
3. $\frac{35!}{30!}$
4. $\frac{88!}{87!}$
5. $\frac{72!}{70!}$
6. $\frac{65!}{62!3!}$
7. ${}_8P_2$
8. ${}_{15}P_0$
9. ${}_9P_9$
10. ${}_{12}C_4$
11. ${}_5C_0$
12. ${}_{32}C_{32}$

Solve the following problems.

13. How many ways can you arrange the letters from the word 'K A R E N'?

14. How many ways can you arrange the letters from the word 'K A R E N' if you want the arrangement to begin with a vowel?

15. All standard license plates in Alaska start with three letters followed by three digits. If repetition is allowed, how many different license plates are there?

16. For $1.99, The Creamery Ice Cream Parlor will put any three different flavored scoops, out of there 25 flavors of ice cream, into a bowl. How many different 'bowls' are there?

17. Suppose those same three scoops of ice cream are on a cone. Now how many arrangements are there?

18. A normal deck of playing cards contains 52 cards. How many five card poker hands can be made?

19. How many ways are there to make a full house (three of one kind, two of another)?

20. What is the probability of getting a full house (three of one kind of card and two of another)? Assume a standard deck and no wild cards.

A bag contains 36 marbles. There are twelve blue marbles, eight red marbles, seven green marbles, five yellow and four white marbles. Without looking you reach into the bag and pull out eight marbles. What is the probability you pull out

21. all blue marbles?

22. four blue and four white marbles?

23. seven green and one yellow marble?

24. at least one red and at least two yellow?

25. no blue marbles?

SAT PREP

1. Matilda remembers only the first four digits of a seven-digit phone number. She is certain that neither of the last three digits is zero. If she dials the first four digits, then dials the last three digits randomly from the non zero digits, what is the probability that she will dial the correct number?

2. Let a Δ b be defined as $\frac{1}{a} + b$ where $a \neq 0$. What is the value of 6 Δ 7?

3. If $4xy + 1 = 1$, what is the value of xy?

4. Eight consecutive integers are arranged in order from smallest to the largest. If the sum of the first four integers is 206, what is the sum of the last four integers?

5. If the points A(4, 1), B(4, 8), and C(-3, 8) form the vertices of a triangle, what is the area of the triangle?

UNIT 12

Closest to the Pin
DESCRIPTIVE STATISTICS

UNIT 12
DESCRIPTIVE STATISTICS

You may not know it but you have been dealing with statistics all your life. You hear them on TV ("Four out of five dentists surveyed ..."), see them in sports (" ... an ERA of .287"), and live with them at school (" ... scored 73% on a test with an average of 60%"). In this unit you finally get to find out what all of this means. Let's start with a straight-forward problem.

1. Mr. Millhouse has 18 students in his gifted and talented science class. On their last exam, the scores were:

92, 97, 80, 94, 100, 100, 99, 91, 93, 100, 95, 90, 96, 98, 99, 95, 96, 100

a) What is the mean of this data set?

b) What is the median of this data set?

c) Does this set contain any outliers? If so, what are they?

Solution.
In part (a), we are asked to find the mean of this data set. The mean is the same as the average and you probably already know how to find that. It is very easy, however, to make a mistake when finding the average because we usually have to sum up so many numbers. Are you really sure you have included them all? That is one reason why it is nice to be able to use a graphing calculator to help you. You can enter the data (check with your teacher and/or your manual to find out how to do this!) and review it to see that you have indeed included all the data points. Then the calculator can do the calculations.

The calculations in finding the mean are not too difficult. We sum all the numbers and then divide by the number of elements in the data set, in this case, 18. Here the sum of all the scores is 1715. To find the mean, $\frac{1715}{18} \approx 95.28$.

The median is the number right in the middle: half the numbers are above the median and half the numbers are below. You can find the median using your calculator as well. If you must find the median by hand, the best way is to list all the data points from highest to lowest, then find the number in the middle. I find the number by counting up from the bottom at the same time (and pace) as I count down from the top. Then I "meet" in the middle. Since there are 18 numbers in this set, there isn't just one number in the middle. The two numbers are the same: 96. That's our median. What do we do if the two numbers are different, say 96 and 95? We average them for a median of 95.5. Note that the median, 96, is very close to the median, 95.28. This often happens, but not always. Be aware that they don't have to be close at all.

100
100
100
100
99
99
98
97
96
96
95
95
94
93
92
91
90
80

Part (c) asks if the data set has any outliers. Outliers are data points that don't quite fit in. Given that these scores were from a gifted and talented class in which the mean is 95.28 and the median is 96, I would say this data set has one outlier: 80. This one point is quite a bit away from the mean, and although it is a respectable score, it does not fit with the rest of them.

Sometimes statistics can tell you a lot about the data. Other times, the statistics might just cloud the situation. It is important when studying or using statistics that you consider more than just what the numbers are. For instance, in this next problem we will consider whether there is bias in the data gathering.

2. The County Board of Education is writing a new graduation exam all students must pass in order to receive a diploma. They have written two different versions of the test and must choose the most appropriate. Both test have the same number of problems and are both scored on a hundred point scale. The Board decides to give the test to two groups of students. The give test number one to Mr. Millhouse's gifted and talented science class. Test number two they give to Mr. Simpson's auto shop class. The scores are:

Test #1 **Mr. Millhouse**	98, 85, 92, 91, 89, 89, 93, 97, 92, 95, 99, 88, 96, 94, 87, 92, 89, 96
Test #2 **Mr. Simpson**	45, 57, 65, 84, 32, 51, 80, 73, 79, 69, 71, 58, 68, 81, 83

a) What is the average for each class?

b) What is the median for each class?

c) What is the range of scores for each test?

d) Is this data unbiased? That is, were the tests given to suitable general classes of students to be able to predict school-wide results? Explain.

Solution.

We calculate the mean and median in the same manner as we did the previous problem. You can use your graphing calculator and should do so for the practice. The mean for Mr. Millhouse's class is approximately 92.3 while the mean for Mr. Simpson's class is 66.4. The median for Mr. Millhouse's class is 92 and for Mr. Simpson's is 69.

Each test had a different range of scores. In Mr. Millhouse's class the scores ranged from a low of 85 to a high of 99. The range for Mr. Simpson's class is a low of 32 to a high of 84.

Now the real question: is the data unbiased? Could you, based on the data here, make a prediction as to how the whole school would do on either test? Mr. Millhouse's class did pretty well on the test. Does that mean the whole school would do well on test #1? Is test #1 too easy for the school? Probably not. Since Mr. Millhouse's class is made up of very bright students (after all, it is a gifted and talented class), we would expect them to do well. But we have no indication of how an average student would do. Since this is a graduation exam we would want all the students who have passed their classes and fulfilled the requirements to pass, not just the brightest. This test could be extremely difficult and Mr. Millhouse's students only did well because they are so bright. Everyone else might fail the test. We just don't know.

Since Mr. Simpson's class did much more poorly on the second test, does that mean test #2 is much more difficult? Probably not. Since Mr. Simpson's class is an auto shop class, these students might not have done their best on the exam. Also, the class could be made up of students not too concerned with graduating high school because they plan on taking jobs as auto mechanics soon. Either way, we can not make any guesses how the rest of the school would do on these tests. A better solution would be to randomly select students from the entire student body to take the test. The larger the sample, the better, to try to make the test a true indication of how the whole school will perform.

Sometimes the mean and median are not enough to represent data accurately. These two statistical measurements tell us the average and middle, but as we have seen, the range of the data can also be important. Other methods for measuring the spread of the data are calculating the **Mean Deviation** (see DS-33) and the **Standard Deviation** (DS-36).

$$\text{Mean Deviation} = \frac{\Sigma|(x - \bar{x})|}{n}$$

$$\text{Standard Deviation} = \sqrt{\frac{\Sigma(x - \bar{x})^2}{n}}$$

Be sure you have these formulas, stating what each symbol means, in your Tool Kit. Include an example as well. We will use these new measurements in analyzing the data in this next problem.

3. After grading her last set of Algebra tests, Ms. Nunnemaker records her scores with a stem and leaf plot.

Stem	Leaf
1	6
2	
3	
4	2
5	1
6	5 7
7	2 7 3
8	1 9 8 8 4 5
9	6 4 3
10	

a) How many data points are represented in the plot?

b) What are the data points?

c) What is the mean and median for this data?

d) Calculate the mean deviation and the standard deviation and explain what they tell us.

e) How do the deviations help to characterize the data better than just the mean and median? Explain.

Solution.
A stem and leaf plot is a quick way to write down all the points to form a "sketch" of a bar graph. If a test were worth from one to 100 points (as in this problem), the numbers one through ten are written down in a column representing the <u>tens</u> digit of the possible score. (Note: the number 10 represents the tens and hundreds digit.) If Ms. Nunnemaker has a score of 72, she would record the ones digit (2) in the "7" row. A score of 51 would be recorded as a "1" in the "5" row. Therefore, each digit to the right of the bar represents a different score. Counting these up we find there are seventeen data points represented here. We can list the data points by combining the tens digit with each ones digit.

16, 42, 51, 65, 67, 72, 77, 73, 81, 89, 88, 88, 84, 85, 96, 94, 93

As you can see, this type of plot offers a great deal of information. Just by looking at the plot we can see that most students scored in the eighties.

Part (c) asks for the mean and median. Using my calculator to be sure I don't make a mistake, I find the mean to be approximately 74.17 and the median is 81.

You should be able to calculate the deviations of part (d) with your graphing calculator. Check with your teacher to be sure it is permissible to do this. If it is, be sure you have the steps written in your Tool Kit. On some graphing calculators (like my older model!) you may have to use tables to calculate the deviations. But, even using tables is easier than calculating these by hand! I will show how to calculate these by hand for those of you who can't use a graphing calculator. The process is LONG and there are many places to make mistakes. Go slowly and carefully!

For the mean deviation, we first must take the absolute value of the difference between each data point, x, and the mean. This means we calculate $|16 - 74.17| = 58.17$, $|42 - 74.17| = 32.17$, $|51 - 74.17| = 23.17$, ... $|96 - 74.17| = 21.83$. We must do this for EACH data point in the set. The sigma notation, Σ, tells us to sum (add) all of these up: $58.17 + 32.17 + 23.17 + ... + 21.83 = 266.47$. Once we have the sum, we divide by n, the number of data points. Therefore, the mean deviation is $\frac{266.47}{17} \approx 15.67$.

Similarly, to calculate the standard deviation, we find the difference between each data point and the mean, this time squaring the result: $(16 - 74.17)^2 \approx 3384.5$, $(42 - 74.17)^2 \approx 1035.3$, ... , $(93 - 74.17)^2 \approx 354.33$.

Next, sum these: $\Sigma(x - \bar{x})^2 \approx 3384.5 + 1035.3 + ... + 354.33 = 7192.47$, and divide by

$n = 17$: $\frac{\Sigma(x - \bar{x})^2}{n} \approx \frac{7192.47}{17} \approx 423.09$. There is one last step with the standard deviation: take the square root: $\sqrt{423.09} \approx 20.57$.

What do these deviations tell us? They give us an idea of how close all the data points are to the mean. For instance, very small deviations tell us all the numbers are very close to the mean. Here, the data is not all bunched up around the mean, but somewhat spread out. Also, if we take the mean and add and subtract one deviation, that will give us some "bounds" on the data.

$\bar{x} \pm$ mean deviation:	74.17 ± 15.67:	58.5 and 89.84
$\bar{x} \pm$ standard deviation:	74.17 ± 20.57:	53.6 and 94.74.

Adding or subtracting two deviations gives even better bounds.

$\bar{x} \pm$ 2(mean deviation):	74.17 ± 31.34:	42.81 and 105.51
$\bar{x} \pm$ 2(standard deviation):	74.17 ± 41.14:	33.07 and 115.31.

Note: If we knew that the test had a maximum of 100 points, we could change the upper bounds to 100 in each of these. In any set of data, **at least** 75% of the data will lie within two standard deviations of the mean. One way the deviations help to characterize the data, then, is to help us pick out points that are outliers. If 75% of the data is within two deviations of the mean, a data point even further way could be considered an outlier. The score of 16 is definitely an outlier!

These numbers help us understand the data better than just the mean and median. This next problem checks to see how well you understand this.

4. After grading the first set of exams, Ms. Vingh calculated the following: mean ≈ 85.6, median = 86, and σ ≈ 5.3. After the second exam, she calculated the new statistics for the second exam. This time she found the mean to be approximately 52.6, a median of 42, and a standard deviation of approximately 33.6. What can you say about these two different sets of test? Which one might have had some outliers in the scores?

Solution.
The first test has a mean value very close to the median. This along with the fact that the standard deviation is very small (5.3) tells us the scores are not spread out. They must all be bunched around the mean. In fact, as we read in DS-36, 95% of the data will be within two standard deviations of the mean. For this set that means 95% of the scores are within 75 to 96.2. The percentage could be higher.

On the second exam, however, the mean and median are ten points apart. Also, the standard deviation is much larger, 33.6. This tells us that the scores are spread out. In fact, we can also guess that there might be one or two high scores pulling the mean up. I would guess that the second test most likely had an outlier or two.

In an attempt to characterize data better, statisticians have come up with numerous procedures and formulas. Although you might not believe it, these procedures and formulas are supposed to simplify the statistical information. This is the case with **z-scores** (read DS-59). A z-score simply turns a raw score (or the original score) into a small descriptive number.

5. A test has a mean of 75 and a standard deviation of 20. Convert the following raw scores into z-scores.

a) 85 **b) 50**

Convert the following z-scores to raw scores.

c) 2.25 **d) -.75**

Solution.
To calculate a z-score, we need to figure out how many standard deviations away a score is from the mean. The score of 85 is ten points away from the mean. That's only half a standard deviation ($\frac{10}{20} = .5$) away so the z-score is .5 or $\frac{1}{2}$. What about a score of 50? This is below the mean so it will have a negative z-score. It is 25 away from the mean, and so $\frac{-25}{20} = -1.25$ is the z-score for 50.

To go the other direction, just reverse the process. A z-score of 2.25 means the raw score is 2.25 standard deviations away from the mean. That means it is 2.25(20) = 45 points away from the mean. Since the mean is 75, the raw score would be 75 + 45 = 120. (Note: don't assume that the scores are bounded by 100! That just may not be the case!) A z-score of -.75 converts as follows: -.75(20) = -15, 75 + -15 = 60.

As pointed out in the book, there is a problem with z-scores. Who wants to get a score of -1, or even 1? Yes it tells you something about how you did: a score of 1, for instance, tells you your score is one standard deviation <u>above</u> the mean. That is a very good score! But, would you rush home to tell your mom "Hey! I got a 1 on my math test today!" If you told your mom you got an 85, however, she would probably have a sense of what that score means and that it is a pretty good score. (Note: an 85 in general doesn't have to be a "good" score, necessarily. For instance, if the test were worth 200 points, 85 points would be a dismal score. Or, if the test were worth 100 points but the mean were 95, then 85 isn't looking too good either. Of course, you elect how much statistical information you will give your mom!)

So often times a z-score is transformed to a new score which also tells us information but has a more appealing number. This is done, in fact, with SAT scores. Read DS-90 and DS-91 to see how this is done.

Another descriptive number is the **percentile** (see DS-71). You probably have seen percentiles before, specially since most school related scores are reported this way. A nice graphical way to represent percentiles is with a box and whisker plot. Read DS-73 before trying this next problem.

---

6. After each test, Ms. Paltrow uses a different statistical measurement and/or graph to report the results to the class. Then each student writes a paragraph about how he/she did compared to the class based on the measurement and/or graph given. This time, Ms. Paltrow shows the class the following box and whisker diagram.

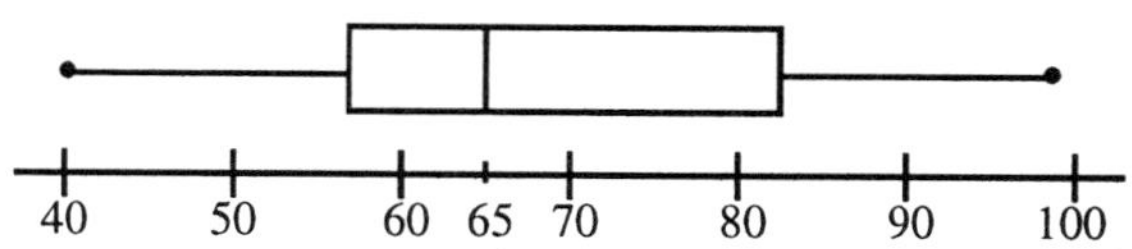

Emmy scored 92 while Howard scored 60. Help each of these students by writing their paragraphs. Include descriptions of how the class did (based on the box and whisker diagram) and how each student did in comparison.

Solution.

Certainly, each person who did this problem would write a different paragraph, however, some key points should be the same. First, from the box and whisker diagram, we see that the high score is approximately 99 while the low score is 40. The median is 65. The 25 percentile is approximately 57; 82 appears to be the 75 percentile. By the shape we can tell that the scores below the median are closer together while the scores above the median are spread out.

Emmy should be very pleased with her performance. She is well above the median and in the highest 25 percentile of the class. This means at least (most likely more) 75% of the students scored below her. Howard should realize that he needs to improve. He scored below the median, with his score landing in the lower quartile (between 25th and 50th percentile). This means the majority of the class did better than he did.

You can expect your teacher to assess your ability to calculate any of the statistical measurements we have discussed (mean, median, standard deviation, etc.) as well has represent information graphically (histogram, stem and leaf plot, box and whisker diagram, etc.). More importantly though, is the ability to interpret the statistical information to gain insight (as in the last example) or to look for bias (see DS-111 as an example). Be prepared to discuss the meaning of the statistical measurements, not just compute them.

One last topic introduced in this unit is the **Binomial Theorem**. The formula written in DS-109, looks very complicated and messy. But, surprisingly, this theorem actually makes your work easier! The Binomial Theorem gives us a method for quickly expanding (multiplying out) binomial expressions. A binomial such as $(x + y)$ is simple enough, but raising it to a power can pose a problem. Most students can square such a binomial easily enough, $(x + y)^2 = x^2 + 2xy + y^2$, but even cubing this same binomial opens up many chances for errors. Most students will do the following:

$$\begin{aligned}(x + y)^3 = (x + y)^2(x + y) &= (x^2 + 2xy + y^2)(x + y) \\ &= x^3 + x^2y + 2x^2y + 2xy^2 + y^2x + y^3 \\ &= x^3 + 3x^2y + 3xy^2 + y^3\end{aligned}$$

To calculate $(x + y)^4$, we would take the last expanded expression for $(x + y)^3$ and multiply by $(x + y)$ again. As you can see, this is a lengthy process!

But, what if I told you I can calculate $(x + y)^7$, for instance, in just two steps? Interested? The Binomial Theorem tells us how.

$$\begin{aligned}(x + y)^7 &= {}_7C_0x^7 + {}_7C_1x^6y + {}_7C_2x^5y^2 + {}_7C_3x^4y^3 + {}_7C_4x^3y^4 + {}_7C_5x^2y^5 + {}_7C_6xy^6 + {}_7C_7y^7 \\ &= x^7 + 7x^6y + 21x^5y^2 + 35x^4y^3 + 35x^3y^4 + 21x^2y^5 + 7xy^6 + y^7\end{aligned}$$

Note: I did use my graphing calculator to calculate ${}_7C_2$ and ${}_7C_3$. So how did I do this? The Binomial Theorem tells us a pattern. When the binomial is expanding, the powers of the first term (x, in this example) decrease while the powers of the second term (y) increase. Also, the first term, ${}_7C_0x^7$, is really ${}_7C_0x^7y^0$ and the last term, ${}_7C_7y^7$, is really ${}_7C_7x^0y^7$. The coefficients of each term are ${}_7C_i$, with i going from zero to seven. Where do all these sevens come from? From $(x + y)^7$: the exponent!

What does $(x + y)^5$ equal? Start by writing out the coefficients. They will all be "five choose" something, starting with "choose zero" and ending with "choose five".

$$(x + y)^5 = {}_5C_0\_\_\_\_ + {}_5C_1\_\_\_\_ + {}_5C_2\_\_\_\_ + {}_5C_3\_\_\_\_ + {}_5C_4\_\_\_\_ + {}_5C_5\_\_\_\_$$

We leave blanks to fill in the powers of x and y. The powers of x decrease from 5 to 0, as the powers of y increase from 0 to 5.

$$(x + y)^5 = {}_5C_0\mathbf{x^5y^0} + {}_5C_1\mathbf{x^4y^1} + {}_5C_2\mathbf{x^3y^2} + {}_5C_3\mathbf{x^2y^3} + {}_5C_4\mathbf{x^1y^4} + {}_5C_5\mathbf{x^0y^5}$$

Now use your calculator to compute ${}_5C_0$, ${}_5C_1$, ${}_5C_2$, ${}_5C_3$, ${}_5C_4$ and ${}_5C_5$. Something to notice and save yourself some work: the coefficients are symmetrical. This means once you have calculated the coefficients half way across the expansion, you have also calculated the rest!

$$(x + y)^5 = x^5 + 5x^4y^1 + 10x^3y^2 + 10x^2y^3 + 5x^1y^4 + y^5$$

Another shortcut to remember: the first coefficient and the last are always one (there is only one way to choose zero objects from a set of n, or n objects from a set of n) and the second coefficient is always n (there are n ways to choose n objects from a set of n).

What should you do if the binomial is not so simple? The same thing, really, with only a minor change. Try the following.

7. Expand the following:

a) $(x + 8)^6$ **b)** $(3x - 2y)^8$

Solution.

Part (a) is slightly different from the examples we have seen above, but we will still expand this in a similar manner. First, write out the coefficients and, by using parentheses as position-holders, the powers of the terms.

$$(x + 8)^6 = {}_6C_0(\ )^6(\ )^0 + {}_6C_1(\ )^5(\ )^1 + {}_6C_2(\ )^4(\ )^2 + {}_6C_3(\ )^3(\ )^3 + {}_6C_4(\ )^2(\ )^4 + {}_6C_5(\ )^1(\ )^5 + {}_6C_6(\ )^0(\ )^6$$

Now you can evaluate the "chooses" as well as fill in each of the terms, x and 8, in the correct position.

$$(x + 8)^6 = {}_6C_0(x)^6(8)^0 + {}_6C_1(x)^5(8)^1 + {}_6C_2(x)^4(8)^2 + {}_6C_3(x)^3(8)^3 + {}_6C_4(x)^2(8)^4 + {}_6C_5(x)^1(8)^5 + {}_6C_6(x)^0(8)^6$$

$$(x + 8)^6 = 1(x)^6(8)^0 + 6(x)^5(8)^1 + 15(x)^4(8)^2 + 20(x)^3(8)^3 + 15(x)^2(8)^4 + 6(x)^1(8)^5 + 1(x)^0(8)^6$$

At this point, check with your teach to see if this needs to be simplified further. Many teachers would be satisfied with this answer (I know I would!). It does take a little more work to simplify this, but only do it if a problem asks you too. To simplify, remember to find the power of 8 first, then multiply by the coefficient.

$$(x + 8)^6 = x^6 + 48x^5 + 960x^4 + 10240x^3 + 61440x^2 + 196608x + 262144$$

For part (b) you want to use "position-holders" as well. Here you have more things to worry about: coefficients for both x and y <u>and</u> a negative sign. Take it slowly!

$$(3x - 2y)^8 = {}_8C_0(\ )^8(\ )^0 + {}_8C_1(\ )^7(\ )^1 + {}_8C_2(\ )^6(\ )^2 + {}_8C_3(\ )^5(\ )^3 + {}_8C_4(\ )^4(\ )^4 + {}_8C_5(\ )^3(\ )^5 + {}_8C_6(\ )^2(\ )^6 + {}_8C_7(\ )^1(\ )^7 + {}_8C_8(\ )^0(\ )^8$$

$$(3x - 2y)^8 = {}_8C_0(3x)^8(-2y)^0 + {}_8C_1(3x)^7(-2y)^1 + {}_8C_2(3x)^6(-2y)^2 + {}_8C_3(3x)^5(-2y)^3 + {}_8C_4(3x)^4(-2y)^4 + {}_8C_5(3x)^3(-2y)^5 + {}_8C_6(3x)^2(-2y)^6 + {}_8C_7(3x)^1(-2y)^7 + {}_8C_8(3x)^0(-2y)^8$$

Remember! If you don't have to, DON'T simplify! If you must, remember to calculate the "chooses" first, then the powers (yes, the coefficients of x and y must also be raised to the exponent!), and then multiply.

$$(3x - 2y)^8 = 1(3x)^8(-2y)^0 + 8(3x)^7(-2y)^1 + 28(3x)^6(-2y)^2 + 56(3x)^5(-2y)^3 + 70(3x)^4(-2y)^4 + 56(3x)^3(-2y)^5 + 28(3x)^2(-2y)^6 + 8(3x)^1(-2y)^7 + 1(3x)^0(-2y)^8$$

$$(3x - 2y)^8 = 6561x^8 + 8(2187x)^7(-2y) + 28(729x^6)(4y^2) + 56(243x^5)(-8y^3) + 70(81x^4)(16y^4) + 56(27x^3)(-32y^5) + 28(9x^2)(64y^6) + 8(3x)(-128y^7) + 256y^8$$

$$(3x - 2y)^8 = 6561x^8 - 34992x^7y + 81648x^6y^2 - 108864x^5y^3 + 90720x^4y^4 - 48384x^3y^5 + 16128x^2y^6 - 3072xy^7 + 256y^8$$

Yuk! Something to notice: since the binomial was the difference of two terms, the signs alternate in the expansion. Again, check with your teacher to find out if simplification is really necessary.

And thus ends Unit 12. The most important thing to remember is that your teacher will expect you to know more than just how to calculate statistical measurements. Your teacher will want you to understand the statistics, interpret the statistics, and recognize problems (such as bias) with the statistics.

UNIT 12
MORE TO TRY

Find the following sums.

1. $\sum_{k=1}^{7} 2k + 1$

2. $\sum_{n=1}^{10} 8\left(\frac{1}{2}\right)^n$

Consider this data set.

430, 800, 750, 320, 680, 540,
550, 620, 570, 600, 710

3. What is the mean?

4. What is the median?

5. What is the standard deviation?

6. Convert these to z-scores.

7. If these are SAT scores for a high school class, what kind of class and/or students do you think they are? Justify your answer.

This year's biology class recently took the AP test. The 23 students' scores are:

3, 4, 4, 3, 5, 4, 5, 5, 4, 3, 3, 2, 5, 4, 2, 3, 3, 4, 2, 3, 4, 5, 2

8. Represent this data in a histogram.

9. What statistical measurement and/or graph would best characterize this data? Explain and justify your choice.

For each of the following situations, decide whether or not you believe the statistics are biased or unbiased. Justify your answer.

10. A recent survey conducted by a research team found that over 80% of the American people knew a great deal about current politics, by answering numerous questions about current federal policies, offices and appointments. The survey was conducted by phoning 2500 people living in Washington, D. C.

11. To determine how many people watch which television shows, the Nilsson Corporation sends out 5,000 surveys to people randomly chosen across the country, being sure every state is represented. If a family is chosen for this survey, they record what they watch on television for one 24 hour period indicated on their survey.

12. To determine the fine motor skills of kindergartners, a group of pediatricians visited 20 randomly selected kindergarten classes and asked the kindergartners to tie their shoes. Based on their results, they concluded that 62% of kindergarten children have adequate find motor skills.

SAT PREP

1. If $0 < t < 1$, which of the following statements must be true?

a) $t^3 < t^2 < t$ b) $t^2 < t^3 < t$ c) $t^2 < t < t^3$

d) $t < t^3 < t^2$ e) $t < t^2 < t^3$

2. Without resting once, Antonia biked for 8 hours. First she rode into the wind, averaging 15 miles per hour, then she turned around and rode the same path with the wind at her back. This time she averaged 28 miles per hour. While hiking, she averaged 2 miles per hour on the way up and 3 miles per hour on the way down. How many total miles did Antonia ride?

a) ≈ 5.2 b) ≈ 10.4 c) ≈ 41.865 d) ≈ 78.14 e) ≈ 156.296

3. A rectangular sheet of paper is folded in half four times. When it is unfolded, how many of the smallest rectangles formed by the creases in the paper, will there be?

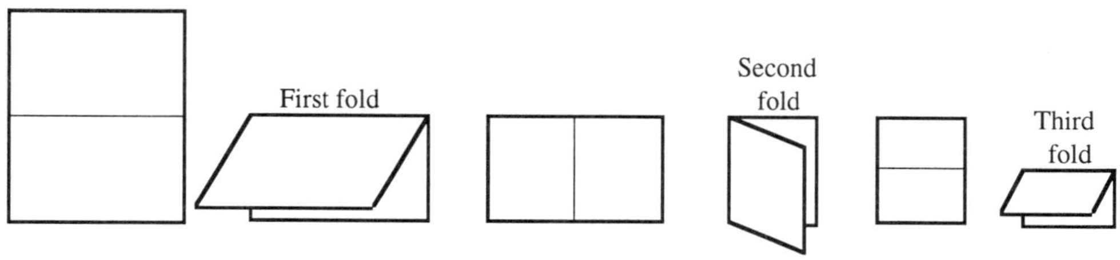

a) 6 b) 8 c) 10 d) 16 e) Cannot be determine from the information given.

4. Tessela can mow the lawn in three hours while it takes Davina four hours to mow the same lawn. How long will it take if they mow the lawn together? (They start from opposite sides working towards each other.)

a) 12 hours b) 7 hours c) ≈ 1.7 hours d) ≈ 42 minutes e) ≈ 9 minutes

5. The sum of two integers is ten while their difference is -6. What are the two numbers?

6. On a certain map, 15 miles is represented by one-half inch. On the same map, how many miles are represented by 3.75 inches?

7. How many of the first one hundred positive integers contain the digit 1?

8. The sum of two consecutive integers is greater than three but less than 13. What is one possible integer fitting these conditions?

9. In the figure at right, the circle is tangent to all four sides of the square. If the diameter of the circle is 1, what is the area of the shaded region?

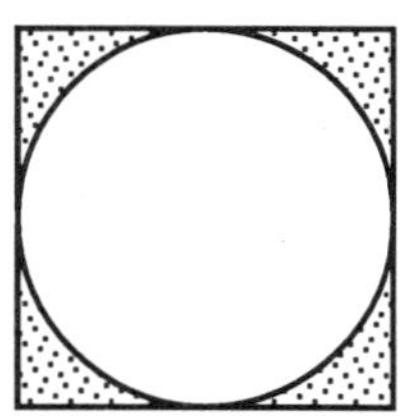

10. Give three values of k for which the trinomial $3x^2 + kx + 6$ factors?

UNIT 13

Return of the Bouncing Ball and Other Topics

REVISITING AND EXTENDING EARLIER TOPICS

UNIT 13
REVISITING AND EXTENDING

This last unit of the book completes our study of Algebra 2. Some of the topics are extensions of ideas you studied earlier in the year. Don't be surprised, then if some what we do in this unit seems familiar and other parts are new ideas.

The first topic takes us back to Unit 2 in which we first encountered sequences. We learned then that a sequence is a function with the domain being the positive integers (1, 2, 3, ...). We studied two types of sequences: arithmetic and geometric. The following are examples of sequences:

(*i*) 2, 4, 6, 8, 10, 12 (*ii*) 4, 7, 10, 13, ..., 604 (*iii*) 3, 6, 12, 24, ...

You should check to see if you can recognize which of the previous sequences are arithmetic and/or geometric.

If we simply make a small change in our sequence, changing each comma to a plus (+), we have just created a **Series**.

(*iv*) 2 + 4 + 6 + 8 + 10 + 12 (*v*) 4 + 7 + 10 + 13 + ... + 604 (*vi*) 3 + 6 + 12 + 24 + ...

Just as with sequences, there are two common types of series: arithmetic series and geometric series. Above, since (*i*) is an arithmetic sequence, the corresponding series (*iv*) is an arithmetic series. Similarly, since (*iii*) is a geometric sequence, (*vi*) is a geometric series. Also, the series (*iv*) and (*v*) are **finite** series since they have a final term while the last series; (*vi*), is **infinite**, meaning the terms continue forever.

When a series is finite, we can always calculate the sum. However, some finite series could be hundreds of terms long. Finding the sum by adding each term could be unbearable. As an example, consider (*v*) above. We don't know how many terms are there, but there are a lot. Would you want to add all of those numbers up? Naturally, this being algebra, there is an easier way to find the sum. Let's try one.

1. Find the sum of the following series.

a) 5 + 10 + 15 + 20 + ... + 100

b) 4 + 7 + 10 + 13 + ... + 604

c) Find the sum of the first hundred odd numbers.

Solution.
In part (a), you might be tempted to start adding the numbers. After all, there are only twenty terms and they are easy to generate. But, even when you know exactly what the terms are and how many there are, it is easy to make a mistake adding a long list of numbers, even with a calculator! RE-1 shows a method for finding the sum of a finite arithmetic series; the box following RE-1 gives the formula. Here is another method similar to the one presented in RE-1. Even if you forget the formula, this method will work every time!

Suppose the sum in part (a) is equal to S. Then

$$S = 5 + 10 + 15 + 20 + ... + 100$$

This sum is still equal to S, even if I reverse the order:

$$S = =100 + 95 + 90 + 85 + ... + 5.$$

Adding these two equations vertically gives,

$$2S = 105 + 105 + 105 + ... + 105$$

There are twenty 105s here. (The sequence is t(n) = 5n. 100 is the 20th term of the sequence.) Therefore, we can write

$$2S = 20(105).$$

But I want to know what S is, not 2S. Solve this for S:

$$\frac{2S}{2} = \frac{20(105)}{2} \quad \text{and} \quad S = 1050$$

As you can see, we needed to know the formula for the nth term of an arithmetic sequence. Let's find that first for part (b).

$$\begin{aligned} t(n) &= 4, 7, 10, 13, ..., 604 \\ t(0) &= 4 \text{ and the common difference is 3.} \\ t(n) &= 3n + 4 \text{ (Check this yourself!)} \\ 604 &= 3n + 4 \\ 600 &= 3n \\ n &= 200 \end{aligned}$$

Is 604 the 200th term in the sequence? Careful! There are 201 terms here because there is also t(0) = 4. Now let's find the sum.

$$\begin{aligned} S &= 4 + 7 + 10 + 13 + ... + 601 + 604 \\ \underline{S} &= \underline{604 + 601 + 598 + 595 + ... + 7 + 4} \\ 2S &= \underbrace{608 + 608 + 608 + 608 + ... + 608 + 608}_{\text{201 of these}} \end{aligned}$$

$$\begin{aligned} 2S &= 201(608) \\ S &= \frac{201(608)}{2} = 61104 \end{aligned}$$

Note: If you prefer the method in RE-1, use it! Or use the formula if you prefer that. Use whatever you need to use to be successful.

You might be thinking that part (c) is not asking for the sum of a series, but it is. The sum that part (c) is asking for is S = 1 + 3 + 5 + ... out to the 100th odd number. In this problem we know there are 100 terms, but we don't know what the 100th term is. You might be able to reason this out (Half the numbers from 1 - 100 are odd, that's 50 odd numbers, so the 100th odd number would be 199.), but I'll use the formulas and methods we have developed to reinforce and practice what we have learned. For this sequence, t(0) = 1 and the common difference is 2. Therefore, t(n) = 2n + 1. (Check: t(0) = 1, t(1) = 3, t(2) = 5.) In this sequence, we will have 100 odd numbers at the term t(99). (Don't forget about t(0)!) So, t(99) = 2(99) + 1 = 198 + 1 = 199. Now we can find the sum.

$$S = 1 + 3 + 5 + ... + 197 + 199$$

$$\begin{aligned} S &= 199 + 197 + 195 + \ldots + 3 + 1 \\ 2S &= 200 + 200 + 200 + \ldots + 200 + 200 \end{aligned}$$

$$\underbrace{\qquad\qquad\qquad\qquad}_{\text{100 of these}}$$

$$\begin{aligned} 2S &= 100(200) \\ S &= \frac{100(200)}{2} = 10000 \end{aligned}$$

Note: I once asked this same question on a test and many students assumed they did something wrong because the sum is even. If you add an even number of odd numbers, the sum will always be even. (Try it!) Adding an odd number of odd numbers will given an odd sum. (Try that too!)

Of course, if we can sum arithmetic series, we can also sum geometric series. Read RE-9 through RE-12 and RE-19 before trying the following.

2. Find the following sums.

a) $2 + 4 + 8 + 16 + \ldots + 4096$

b) $19683 + 6561 + 2187 + \ldots + \frac{1}{3} + \frac{1}{9}$

c) $24 + 12 + 6 + 3 + \ldots$ **(Yes! It is infinite!)**

Solution.

The method outlined in the problems RE-9 through RE-12 is the standard method for finding the sum of a geometric sequence. The formula in the box in RE-12 cuts out all of the intermediate steps of the method. Since I used the method in the last example I will use the formula in this example.

The formula says that the sum of a geometric series is $S = \frac{a(r^n - 1)}{r - 1}$ where r is the common ratio, a is the first term and n is the number of terms we are adding up. In part (a) we can quickly find a (a = 2), and the common ratio r ($\frac{4}{2} = 2$), so all that we need to find is n, the number of terms there are. Fortunately, since we know a and r, we can write a formula for the nth term, set it equal to 4096, and solve for n. Here, $t(n) = ar^n$, remembering that $t(0) = 2$ is our first term. Therefore, $t(n) = 2(2^n)$. Thus, we can write $2(2^n) = 4096$ which is the same as $2^{n+1} = 4096$. You might be able to solve this by guess and check or even inspection. I will use logarithms to show a precise method.

$$\begin{aligned} 2^{n+1} &= 4096 \\ \log(2^{n+1}) &= \log 4096 \\ (n+1)\log 2 &= \log 4096 \\ n + 1 &= \frac{\log 4096}{\log 2} \\ n &= \frac{\log 4096}{\log 2} - 1 \\ n &= 11 \end{aligned}$$

Therefore, $t(11) = 4096$ but that means there are **12** terms listed here because we cannot forget t(0). Now we are ready to use the formula for the sum of a geometric series.

Recall that to solve $a^x = b$; $x = \frac{\log(b)}{\log(a)}$. so then if $a^{n+1} = b$; $n + 1 = \frac{\log(b)}{\log(a)}$.

$$S = \frac{a(r^n - 1)}{r - 1}$$

We will do this same thing for part (b). In part (b), a = 19683 and $r = \frac{1}{3}$. How many terms are listed here? Rather than count them, I will find the formula for the nth term and solve for n. Here, $t(n) = 19683(\frac{1}{3})^n$. Then $\frac{1}{9} = 19683(\frac{1}{3})^n$. This might seem difficult to solve so I will show all steps.

$$\begin{aligned} \frac{1}{9} &= 19683(\tfrac{1}{3})^n \\ \frac{1}{9} &= 19683(\tfrac{1}{3^n}) \\ \frac{3^n}{9} &= 19683 \\ 3^n &= 177147 \\ n &= \frac{\log 177147}{\log 3} \\ n &= 11 \end{aligned}$$

Therefore, $t(11) = \frac{1}{9}$, but once again this means there are **12** terms listed here because we must count t(0) = 19683. Now we can find the sum.

$$\begin{aligned} S &= \frac{a(r^n - 1)}{r - 1} \\ &= \frac{19683((\frac{1}{3})^{12} - 1)}{\frac{1}{3} - 1} \\ &\approx 29524.4 \end{aligned}$$

Once I had the numbers in the formula, I used my calculator to calculate this messy fraction. Keep this in mind: we are finding a sum so it is okay to get rational (decimal) numbers. Also, on some problems you might calculate the sum of a series to be negative. That is okay if you are adding up a fair amount of negative numbers. Here, where each term is positive, we should not (better not!) get a negative sum.

For part (c) you might be wondering how we can add up an infinite number of terms. Won't the sum just keep getting bigger and bigger? It turns out that as long as the common ratio is less than one, the sum is finite and can be found. And, surprisingly, the sum is very easy to find. All we need to know is a, the initial term and r, the common ratio. Therefore this sum is

$$S = \frac{a}{1 - r} = \frac{24}{1 - \frac{1}{2}} = \frac{24}{\frac{1}{2}} = 48$$

The next topic presented in this unit is **Mathematical Induction**. This topic might perplex you; the example presented in the text does not look like any proof you might have seen before. But, mathematical induction is a very algorithmic sort of proof. In that sense, students oftentimes find it very easy to master. Read through RE-26 before trying this next problem.

3. Prove that $\frac{1}{1\cdot 2} + \frac{1}{2\cdot 3} + \frac{1}{3\cdot 4} + \ldots + \frac{1}{n\cdot(n+1)} = \frac{n}{n+1}$ **by mathematical induction.**

Solution.

Almost all proofs by induction are done the same way. First you show the statement is true for n = 1. Then you assume it true for n = k and show that this implies it true for n = k + 1. So that is our outline for our proof.

First, we show it true for n = 1. $\frac{1}{1\cdot 2} = \frac{1}{2}$ Check! This is true!

Next, we assume the statement is true for n = k. Always write this statement out, realizing that you are saying that it is true. NEVER skip this step. It is usually crucial to being able to prove it true for n = k + 1. For n = k we have:

$$\frac{1}{1\cdot 2} + \frac{1}{2\cdot 3} + \frac{1}{3\cdot 4} + \ldots + \frac{1}{k\cdot(k+1)} = \frac{k}{k+1} \quad \text{(A)}$$

Remember: that statement (A) above is **true**! Next we want to prove it true for n = k + 1. To start, right out the left side of this statement for n = k + 1.

$$\frac{1}{1\cdot 2} + \frac{1}{2\cdot 3} + \frac{1}{3\cdot 4} + \ldots + \frac{1}{k\cdot(k+1)} + \frac{1}{(k+1)\cdot(k+2)} \quad \text{(B)}$$

We want that last statement, (B), to equal $\frac{k+1}{k+2}$. But look closely at (B). We already know what the first k terms equal. The first k terms are just the left side of equation (A). Therefore, we can replace these k terms with what they equal: $\frac{k}{k+1}$. Therefore,

$$\frac{1}{1\cdot 2} + \frac{1}{2\cdot 3} + \frac{1}{3\cdot 4} + \ldots + \frac{1}{k\cdot(k+1)} + \frac{1}{(k+1)\cdot(k+2)} = \frac{k}{k+1} + \frac{1}{(k+1)(k+2)}$$

We are not quite at the form we are looking for. But, as in most induction proofs, now we do algebra to make this look different. And when you have two fractions in an addition problem, the best thing to do is add them! To do that we need common denominators.

$$\begin{aligned}\frac{k}{k+1} + \frac{1}{(k+1)(k+2)} &= \frac{k}{k+1}\cdot\frac{(k+2)}{(k+2)} + \frac{1}{(k+1)(k+2)} \\ &= \frac{k(k+2)}{(k+1)(k+2)} + \frac{1}{(k+1)(k+2)} \\ &= \frac{k(k+2)+1}{(k+1)(k+2)} \\ &= \frac{k^2+2k+1}{(k+1)(k+2)} \\ &= \frac{(k+1)(k+1)}{(k+1)(k+2)} \\ &= \frac{k+1}{k+2}\end{aligned}$$

which was what we wanted! This means we have proved the statement by induction.

An important point to remember when doing these kinds of proofs is the basic routine: show the statement is true for n = 1, assume it is true for n = k, and use algebra to "derive" it true for n = k + 1.

The next part of the unit covers conic sections, or general quadratics. A **general quadratic equation** is any equation of the form $ax^2 + by^2 + cx + dy + e = 0$ where both $a \neq 0$ and $b \neq 0$. (Note: a general quadratic equation can also contain an xy term, but in this book we ignore this case.) Depending on what the coefficients are, this equation can represent a parabola, a circle, an ellipse or a hyperbola. When you are given a general quadratic equation, the key is in figuring out which type it is. It might be best to keep a chart in your Tool Kit explaining how to recognize each one based on the problems in this unit. Do that, then try the following.

4. For each of the following equations, write the equation in graphing form, identify the curve, and list the following information: center, focus or foci, asymptotes and eccentricity.

a) $9x^2 + 4y^2 - 18x + 24y + 9 = 0$

b) $x^2 + y^2 - 8x + 14y - 35 = 0$

c) $16y^2 - 36x^2 - 160y - 144x - 320 = 0$

Solution.

To write general quadratic equations in graphing form we will complete the square, once for each variable. Once we do this, we will be able to recognize which curve each equation represents. To complete the square, I'll first bring like variables together. So in part (a),

$$9x^2 + 4y^2 - 18x + 24y + 9 = 0$$
$$9x^2 - 18x + 4y^2 + 24y = -9$$

Next, factor out the coefficients of the squared terms and complete the square:

$$9(x^2 - 2x + \quad) + 4(y^2 + 6y + \quad) = -9$$
$$9(x^2 - 2x + \mathbf{1}) + 4(y^2 + 6y + \mathbf{9}) = -9 + \mathbf{9} + \mathbf{36}$$

Note: the step above is probably the most difficult part. We completed the square by filling in the necessary constant so that the trinomial is a perfect square. In the first trinomial I added a **1** within the parentheses while in the second trinomial I added a **9**. But, these aren't really a **1** or a **9** because the distributive rule applies. If we distributed the 9 and 4 in front of the parentheses we would see that we really added a **9** and a **36**. Therefore we must add **9** and **36** to the other side as well to keep the equality intact. Let's continue.

$$9(x - 1)^2 + 4(y + 3)^2 = 36$$
$$\frac{9(x - 1)^2}{36} + \frac{4(y + 3)^2}{36} = \frac{36}{36}$$
$$\frac{(x - 1)^2}{4} + \frac{(y + 3)^2}{9} = 1$$

Now we can see that this is an ellipse with center (1, -3). In an ellipse, the foci lie along the longer axis, which is the vertical axis in this problem. (You can recognize this quickly because the larger number is under the expression containing y. If the larger number had been under the x, the longer axis would have been horizontal.) The distance to the foci is $c = \sqrt{9 - 4} = \sqrt{5}$ (the 9 and the 4 coming from the denominators above). Since the foci are on the vertical axis at a distance of $\sqrt{5}$ from the center, the coordinates of the foci are $(1, -3 \pm \sqrt{5})$. The eccentricity, e, is found by $e = \frac{c}{a} = \frac{\sqrt{5}}{3}$.

It is best to record the information and formulas in your Tool Kit and use them rather than trying to memorize this. It can be very confusing! The general equation for an ellipse is $\frac{(x - h)^2}{a^2} + \frac{(y - k)^2}{b^2} = 1$ with the center at (h, k). When the longer axis, the major axis, is horizontal, the denominator under the x term is larger. When the major axis is vertical we normally write it as $\frac{(x - h)^2}{b^2} + \frac{(y - k)^2}{a^2} = 1$. In either case, you find the distance to the foci is c where $c^2 = a^2 - b^2$ and the eccentricity is $e = \frac{c}{a}$. All of this should be in your Tool Kit!

Let's try part (b), beginning with the same algebra.

$$\begin{aligned} x^2 + y^2 - 8x + 14y - 35 &= 0 \\ x^2 - 8x + y^2 + 14y &= 35 \\ (x^2 - 8x + 16) + (y^2 + 14y + 49) &= 35 + 16 + 49 \\ (x - 4)^2 + (y + 7)^2 &= 100 \end{aligned}$$

This is the equation of a circle with center (4, -7) and radius 10. There is no foci since it is a circle (c = 0), and so the eccentricity is 0 ($e = \frac{0}{10}$). Note: if you accidentally kept going, dividing both sides by 100 next, don't worry. It can be written that way and you would have gotten the same c and e values. You would have to "undo" your work to determine the radius however. The key to recognizing that this is a circle is noticing that neither of the squared terms has coefficients.

Let's do the same algebra for part (c).

$$\begin{aligned} 16y^2 - 36x^2 - 160y - 144x - 320 &= 0 \\ 16y^2 - 160y - 36x^2 - 144x &= 320 \\ 16(y^2 - 10y + \quad) - 36(x^2 + 4x + \quad) &= 320 \\ 16(y^2 - 10y + 25) - 36(x^2 + 4x + 4) &= 320 + 400 - 144 \end{aligned}$$

Note: When I factored the coefficient of x^2 out, I took out **- 36.** That left a "**+ 4x**". Also, remember that when I completed the square, I really didn't add a 25 to the left, I added a 16(25) when the distributive rule is applied. Similarly, I didn't really add a 4 in the second trinomial. What I really added was -36(4) = -144. Let's continue.

$$\begin{aligned} 16(y - 5)^2 - 36(x + 2)^2 &= 576 \\ \frac{16(y - 5)^2}{576} - \frac{36(x + 2)^2}{576} &= \frac{576}{576} \\ \frac{(y - 5)^2}{36} - \frac{(x + 2)^2}{16} &= 1 \end{aligned}$$

Since this is a difference of two expressions, this is a hyperbola. Also, this hyperbola "sits" vertically because the y variable is the positive term. The "center" of this graph is at (-2, 5). The vertices are at $(-2, 5 \pm 6)$, in other words, at (-2, 11) and (-2, -1). (The 6 in "5 ± 6" came from $\sqrt{36}$, the square root of the first denominator.) The distance to the foci is $c = \sqrt{36 + 16} = \sqrt{52} = 2\sqrt{13}$. This means the foci are at $(-2, 11 + 2\sqrt{13})$ and $(-2, -1 - 2\sqrt{13})$. (Remember: c is the distance from the <u>vertex</u> to the focus.) The eccentricity is $e = \frac{2\sqrt{13}}{6} = \frac{\sqrt{13}}{3}$. Since this is a hyperbola, we do need to find the equations for the asymptotes. To do this we need two things: we need to know their slopes and we need to know a point they pass through. The point is easy: both asymptotes go through the "center", which in this case is (-2, 5). The slope is $\pm\frac{6}{4} = \pm\frac{3}{2}$. Therefore, the equations are $y = \frac{3}{2}x + 8$ and $y = -\frac{3}{2}x + 2$. Check these yourself by using the point and the slopes.

All the information about hyperbola can be found in the box of RE-75. And, one last reminder: keep all information about general quadratic equations in your Tool Kit! There is too much information to try to memorize.

The last topic in this unit is **Dividing Polynomials**. The method outlined in RE-81 is probably the best method for dividing polynomials (I say "best" because I notice I make fewer mistakes when using that method!), but there is another common method. I will show it in the next example. You can use this method or the rectangle method of RE-81, whichever you prefer.

5. Divide

a) $x^3 + 4x^2 - 7x - 10$ **by** $x + 1$ **b)** $3x^3 - 5x^2 - 34x + 24$ **by** $3x - 2$

Solution.

For part (a), I will present the alternate method of dividing I mentioned. This method is really the long division you learned back in fourth grade. For that reason, some students feel very comfortable using it. To see how it works, let's first divide **by hand** 7452 by 23.

$$23\overline{)7452}$$

$$\begin{array}{r} 3 \\ 23\overline{)7452} \\ \underline{69} \\ 5 \end{array}$$

When doing such a division without a calculator, we first made an educated guess as to how many times 23 went into 74; we just looked at the "front" of the number, temporarily ignoring the rest. (Do this problem along with me so that you will remember the steps.) Once we guessed three times, we wrote 3 above the four in 74. Do you remember what we did next? We multiplied 3 times 23, writing the answer under the 74. Since 3 times 23 is 69, I wrote the 69 under the 74. Next, we subtracted, leaving 5.

$$\begin{array}{r} 32 \\ 23\overline{)7452} \\ \underline{69} \\ 55 \\ \underline{46} \\ 9 \end{array}$$

This is probably coming back to you, I hope. Once we subtracted, we brought the next digit, in this case 5, down. Now we ask how many times 23 goes into 55? We make an educated guess again, of 2 times. We write the 2 above the 5 and again we multiply 2 times 23. This answer we write below the 55 and once again, we subtract. This leaves 9.

$$\begin{array}{r} 324 \\ 23\overline{)7452} \\ \underline{69} \\ 55 \\ \underline{46} \\ 92 \\ \underline{92} \\ 0 \end{array}$$

The process is repeated until finished: make a guess, record it , multiply it, subtract, bring down the next digit. So, we bring down the 2, write it next to the 9, and guess how many times 23 goes into 92. This time it goes in exactly 4 times. We multiply 4 times 23, and write the answer below the 92. We subtract and get 0. Therefore, 23 goes into 7452 exactly 324 times, no remainders.

Now we can do the same thing with polynomials. We will write it out as a long division problem and look at only part of the polynomial, temporarily ignoring the rest. We will follow the same pattern: guess, multiply, subtract and bring down.

$$x+1\overline{\big)\,x^3+4x^2-7x-10}$$

Start by writing the problem as a long division problem. Keep the polynomials in descending order of the powers and, if a term is missing, be sure to include it anyway. For instance, if the polynomial has no x^2 term, still include it in the descending order of the exponents as $0x^2$.

$$\begin{array}{r} x^2 \\ x+1\overline{\big)\,x^3+4x^2-7x-10} \end{array}$$

Then, just as the long division, we need to guess how many times $x + 1$ goes into $x^3 + 4x^2$. All we really need to do here is figure out how many times x goes into x^3. As before, we just look at the first terms, temporarily ignoring the rest. Since x goes into x^3, x^2 times, we write x^2 above the $4x^2$. As before we multiply x^2 times $x + 1$ and write it below. Then subtract. Be careful when subtracting. That's where most people make mistakes.

$$\begin{array}{rl} & x^2 \\ x+1\big) & \overline{x^3+4x^2-7x-10} \\ & \underline{x^3+x^2} \\ & 3x^2 \end{array}$$

$$\begin{array}{rl} & x^2+3x \\ x+1\big) & \overline{x^3+4x^2-7x-10} \\ & \underline{x^3+x^2} \\ & 3x^2-7x \end{array}$$

Once we subtract, we bring down the next term: $-7x$. How many times does x go into $3x^2$? $3x$ times. Write this above the $-7x$. Multiply $3x$ times $x + 1$ and then subtract. Note: when multiplying, don't forget about the distributive rule. You must multiply $3x(x + 1) = 3x^2 + 3x$.

$$\begin{array}{rl} & x^2+3x \\ x+1\big) & \overline{x^3+4x^2-7x-10} \\ & \underline{x^3+x^2} \\ & 3x^2-7x \\ & \underline{3x^2+3x} \\ & -10x \end{array}$$

Here is a typical example of where someone will make a mistake when subtracting. Subtract straight down: $3x^2 - 3x^2 = 0$ and $-7x - 3x = -10x$.

$$\begin{array}{rl} & x^2+3x-10 \\ x+1\big) & \overline{x^3+4x^2-7x-10} \\ & \underline{x^3+x^2} \\ & 3x^2-7x \\ & \underline{3x^2+3x} \\ & -10x-10 \\ & \underline{-10x-10} \\ & 0 \end{array}$$

One last time: x goes into $-10x$, -10 times. Write this above the -10. Multiply -10 times $x + 1$ and subtract. Since we got zero here, this tells us that $x + 1$ goes into this polynomial of degree three evenly (no remainder). If there were a remainder, there are different ways to handle it. One way is to simply write "remainder ___" and state the remainder. Another is to write the remainder as a fraction. Check with your teacher to see which method is preferred.

This problem could have been done (and perhaps even more easily) with rectangles. Use the method you prefer.

Since I used long division for part (a), I will use rectangles for part (b). With rectangles I like to start with a "non-specific" rectangle; that is, I don't complete the rectangle since I don't know how big it will be initially . I write the divisor 3x - 2 across the top. Then I fill in the term with the largest exponent in the upper left corner. Given a rectangle with an area of $3x^3$ and with one side length of 3x, what must the other side length be? It must be x^2. We fill this in on the left side.

	$3x$	-2
	$3x^3$	

Multiply by x^2 by -2 to fill in the area of the rectangle in the upper right: $-2x^2$. This next part is the trickiest. Look back at our dividend (the dividend is the polynomial we are dividing <u>into</u>: $3x^3 - 5x^2 - 34x + 24$). We need to have total of $-5x^2$. Right now there are only $-2x^2$ in the rectangle. We need to have $-3x^2$ more. Add this into the next rectangle on the left.

	$3x$	-2
x^2	$3x^3$	$-2x^2$

	$3x$	-2
x^2	$3x^3$	$-2x^2$
	$-3x^2$	

Now repeat the process. This new rectangle has an area of $-3x^2$ and the length of the one side is 3x. What must the length of the other side be? It must be -x. Fill this in on the left and multiply to find the area of the rectangle on the right: 2x.

	$3x$	-2
x^2	$3x^3$	$-2x^2$
$-x$	$-3x^2$	$2x$

Now we look back at our dividend to see how many x's we should have. -34x. That means we need to put -36x into the rectangle ($-36x + 2x = -34x$, which is what we really want.) With this area filled in we ask ourselves what the length of the other side must be if the area is -36x and one side is length 3x. -12. This leaves the last rectangle to have area 24, which is what we want.

	$3x$	-2
x^2	$3x^3$	$-2x^2$
$-x$	$-3x^2$	$2x$
-12	$-36x$	24

Therefore, $3x^3 - 5x^2 - 34x + 24$ divided by $x + 1$ is $x^2 - x - 12$.

Note: when you are doing these problems you do not need to redraw the rectangle again and again as I did. Since division is a dynamic process, I am just trying to illustrate this by redrawing the diagram to show the progression.

This unit has a lot in it! Be sure you know what is expected from you for class and for your teacher. Include all necessary information on your Tool Kit. And, if you find that you need and/or want more practice on these topics, you can find them in other Algebra 2 books. Ask your teacher is he/she has a book you can borrow. Most older math text books will have many problems on series, general quadratic equations and dividing polynomials. You may need to consult a higher level math book (community college algebra or pre-calculus) to find lots of mathematical induction problems. Several examples of all of these are provided for you in the More To Try section.

Good luck! I hope your year in Algebra 2 has been fun and knowledgeable.

UNIT 13
MORE TO TRY

Find the following sums.

1. $9 + 12 + 15 + \ldots + 54$
2. $1 + 4 + 16 + \ldots + 16384$
3. $3072 + 768 + 192 + \ldots + \frac{3}{16}$
4. $6 + 2 + -2 + \ldots + -50$
5. Sum of the first 200 positive even numbers.
6. $162 + 54 + 18 + \ldots$

Prove the following by induction.

7. $2n \le 2^n$ for all natural numbers n.
8. $\frac{1}{2^1} + \frac{1}{2^2} + \frac{1}{2^3} + \ldots + \frac{1}{2^n} = 1 - \frac{1}{2^n}$
9. $2^n \ge 2n + 1$ for $n \ge 3$.

For each of the following, write the equation in graphing form, identify the curve, and list the following information: center, focus or foci, asymptotes, and eccentricity.

10. $x^2 + 4y^2 + 4x + 8y - 8 = 0$
11. $16y^2 - 25x^2 - 192y - 150x - 49 = 0$
12. $3x^2 - 12x - y + 19 = 0$
13. $x^2 + y^2 - 12x + 10y - 108 = 0$

SAT PREP

1. $(5 + 6)^2 = ?$

   a) $(2 \times 5) + (2 \times 6)$ b) $5^2 + 6^2$ c) 61

   d) 11^2 e) $5^2 \times 6^2$

2. If $6x - 7y = 12$, what is the value of $-2(6x - 7y)$?

   a) 42 b) 13 c) -1 d) -24 e) -42

3. The average (arithmetic mean) of three numbers is 25. If two of the numbers are 25 and 30, what is the third number?

   a) 35 b) 30 c) 25 d) 20 e) 15

4. People from the country of Turpa measure with different units. Each curd is 7 garlongs long and each garlong is made up of 15 bleebs. How many complete curds are there in 510 bleebs?

   a) 105 b) 15 c) 5 d) 4 e) 2

5. If $x^2 - y^2 = 12$ and $x - y = 2$, what is the value of $x + y$?

6. Five consecutive integers sum to 25. What is the largest of these consecutive numbers?

7. For all positive integers m and n, we define m ‡ n to be the whole number remainder when m is divided by n. If 11 ‡ n = 3, what does n equal?

8. At Pies R Us, each pie is cut into a slice as shown in the figure at right. Each slice of pie has a central angle of 30°. They sell the pies by the slice. If the weight of each pie is uniformly distributed, weighing 108 grams, how much does each slice weigh, in grams?

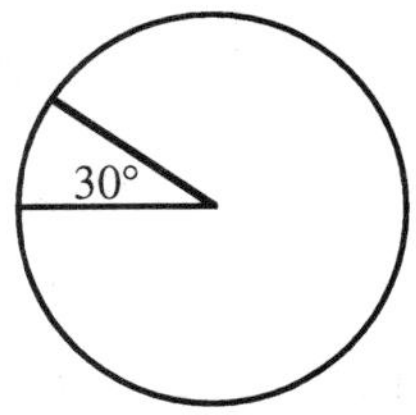

9. In the figure at right, what is the area of the shaded region if that region is a square?

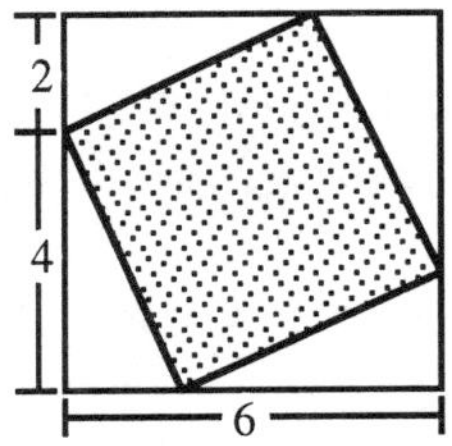

10. What is the sixth term in the sequence beginning 432, 72, 12, ... ?

ANSWERS & HINTS

Unit 1, More to Try

1. $x = \frac{-45}{7}$ **2.** x = 3 **3.** $x \approx 2.97$
4. x = 2, y = 6 **5.** x = -21, y = -5
6. $x = \frac{-8}{5}$ **7.** $x = \frac{c - b}{a}$ **8.** $x = \frac{c}{b} + a$ or

$x = \frac{c + ab}{b}$ **9.** $x = \frac{3 \pm 2\sqrt{3}}{3}$ **10.** x = -3, 7
(Did you remember to set equal to zero first?) **11.** $x = \frac{1}{4}, \frac{8}{3}$ **12.** $x = \pm 4\sqrt{2}$

13. -2 times -3 is 6, not 5, $x = \frac{15}{7}$

14. Didn't set equal to zero, $x = \frac{-3}{2}$, 1

15.

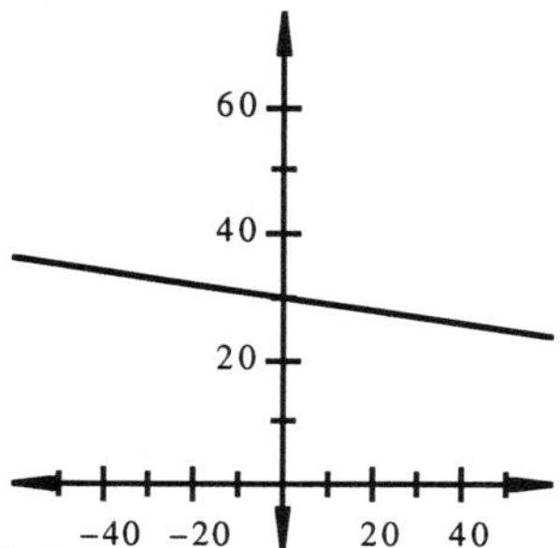

16.

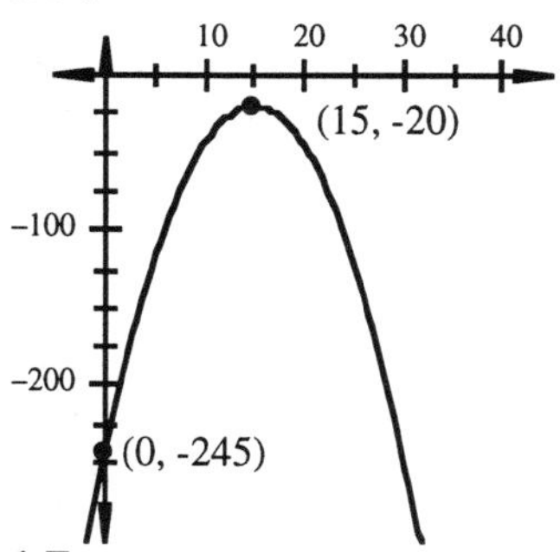

17.

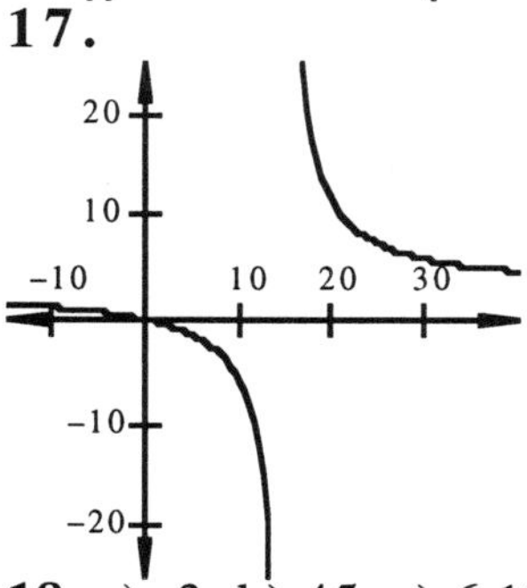

18. a) -3 b) 45 c) 6.1875 **19.** a) 25.8 b) .888 c) 2035.8 **20.** a) $\frac{-8}{3}$ b) 1.8 c) not possible (dividing by zero is undefined or not possible.) **21.** The graph shows the cost for various # of gallons of milk for two different companies. **22.** Point A is where the two graphs cross so it lies on both lines. This is the point (2, 5.23). This means from both companies, two gallons of milk will cost $5.23, the same amount.
23. For Moo For You, the slope is 1.99 and that is the price in dollars per gallon of milk. For Farm Fresh, the slope is 2.39 and that is their price of milk per gallon.
24. The y-intercept for Moo For You is 1.25 and for Farm Fresh it is 0.45. Since this is the cost for zero gallons of milk, this probably represents the cost of the container the milk comes in. **25.** This graph shows the number of gallons of gas a Honda Accord and a Geo Prism use as the drive.
26. From the graph we know that the Honda Accord has a 16 gallon gas tank (the y-intercept, since the tank would be full when it has driven zero miles), has a 352 mile driving range (the x-intercept, because at that point, y is zero so there is no more gas), and it goes 22 miles for each gallon of gas (the slope), or in other words, gets 22 mpg. **27.** The Geo Prism has a 10 gallon gas tank (y-intercept), a range of 400 miles (x-intercept), and gets 40 miles per gallon.
28. It does not make sense to extend these lines. Extending the lines into the second quadrant would imply the cars can travel "negative miles" which has no meaning. Extending the lines into the fourth quadrant would imply the cars have "negative gallons" in them, which also does not make sense. **29.** When you investigate a function you write down everything you can about the equation. Here are several. Your teacher might tell you to include more or less. Be sure you know what your teacher wants and have a list on your tool kit. The graph of this is curved, there is no x- or y-intercept, it is a function, the domain (acceptable x values) is $x \geq 6$, the range (possible y values) is $y \geq 1$. The vertex (or "starting point") is (6,1). **30.** This graph is curved, it has two unconnected parts, it has no x-intercept, the y-intercept is $(0, -\frac{1}{6})$, the domain is all x except 6 and the range is all y except 0. (This next one you probably don't know yet, but remember it for later.) This has a horizontal asymptote at y = 0 and a

vertical asymptote at $x = 6$. This is a function.

SAT Prep

1. D **2.** A **3.** D **4.** D **5.** A **6.** $\frac{1}{2}$

7. 300 **8.** 8 **9.** $\frac{15}{16}$ **10.** $\frac{1}{10}$

Unit 2, More to Try.

1. $7x^6y^{14}$ **2.** $196608x^{21}$ **3.** 648ab **4.** $15625x^{42}a^{18}b^6$ **5.** $-1.5xy^2z^9$ **6.** $-\frac{500b^{12}}{a^9}$ **7.** $4x^8y^7$ **8.** $-486a^{34}x^{20}$ **9.** 2, 7, 12, 17, 22, arithmetic, common difference 5 **10.** 3, -5, -13, -21, -29, arithmetic, common difference -8 **11.** 0, 8, 14, 18, 20, neither **12.** 1, -4, 16, -64, 256, geometric with common ratio -4. **13.** 1, $\frac{1}{4}$, $\frac{1}{16}$, $\frac{1}{64}$, $\frac{1}{256}$, geometric, common ratio $\frac{1}{4}$ **14.** 0, 2, 6, 12, 20, neither **15.** 8, 8, 8, 8, 8, both, common difference 0, common ratio 1 **16.** 1, $\frac{7}{4}$, $\frac{5}{2}$, $\frac{13}{4}$, 4, common difference $\frac{3}{4}$ **17.** geometric (remember to check for common ratios!), $t(n) = 48\left(\frac{1}{2}\right)^n$ **18.** arithmetic, $t(n) = 7n - 4$ **19.** arithmetic, $t(n) = -4n + 43$ **20.** Don't be scared by the fractions! Just look for the common difference and/or ratio. geometric, common ratio:

$\frac{1}{2} \div \frac{2}{3} = \frac{1}{2} \times \frac{3}{2} = \frac{3}{4}$. Check: $\frac{3}{8} \div \frac{1}{2} = \frac{3}{8} \times \frac{2}{1} = \frac{3}{4}$ √.

$t(n) = \frac{2}{3}\left(\frac{1}{2}\right)^n$ **21.** Because this alternates sign, it is not both. No common difference. Geometric, $t(n) = 5(-1)^n$

22. geometric, $t(n) = 10\left(\frac{1}{10}\right)^n$

23.&24.

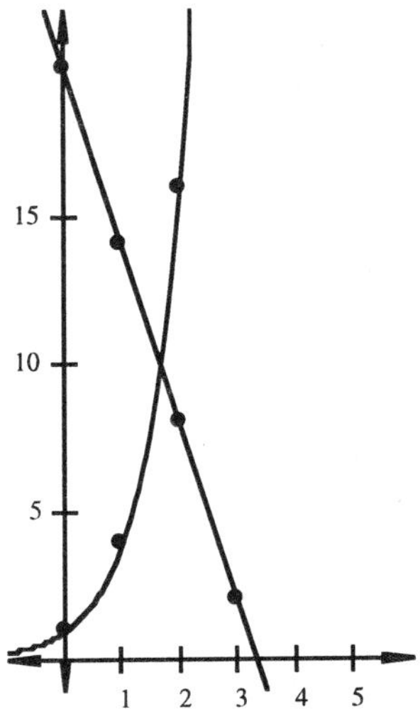

Note: The lines should not be drawn in. Remember that sequences are **discrete functions**. I drew them in just to make the graph clearer. The linear graph is 23. **25.** NO they don't. Although the graphs cross, there is not a point (which represent the terms) in common with both graphs.

26. Geometric, $t(n) = 25\left(\frac{1}{5}\right)^n$

27. Arithmetic, $t(n) = \frac{1}{2}n + 4$ **28.** This is geometric! (Multiply by 1.0005 each time.) In 1998, $54(1.0005)^{450} \approx 67.6$ inches. **29.** Arithmetic, $t(n) = .4n + 5.4$. We want to know when this will equal (or more likely, when will it be more than) 24.99. $0.4n + 5.4 = 24.99$, $n = 48.975$, so in 49 weeks. In 49 weeks he will have $t(49) = 0.4(49) + 5.4 = 25$. That won't cover tax! He will need another 3 or 4 weeks, depending on the tax rate, before he will be able to cover the entire cost. Looks like Davis won't have his race car set till his fifth birthday! **30.** This is a function, it represents an arithmetic sequence, the graph is discrete, but points are linear, the y-intercept is (0, 18), there is no x-intercept. The domain is all whole numbers (0, 1, 2, 3, ...), the range is the sequence itself: 18, 14, 10, 6, 2, -2, ... No asymptotes.

SAT Prep

1. B **2.** C **3.** E **4.** D **5.** E **6.** B **7.** D **8.** C **9.** B **10.** D

Unit 3, More to Try.

1. $\frac{x^{13}}{972}$ **2.** $\frac{3y^8}{x^7}$ **3.** $\frac{x^3}{243y^{16}}$ **4.** 1 (Look at the whole problem before you start with the individual parts!) **5.** -32

6. $8a^6b^9$ **7.** $\frac{ac^{3/4}}{b}$ **8.** $\frac{12}{x^3}$ **9.** $\frac{16}{27}$

10. 13122 **11.** 16 **12.** 4 **13.** -2

14. $-\frac{2}{3}$ **15.** $\frac{5}{11}$ **16.** 2 **17.** 0

18. ≈ 2.91 **19.** 5 **20.** Both sides need to the raised to the $\frac{1}{6}$ power, not divided by 6. x ≈ -4.35 **21.** Since 5 and 10 cannot be written as the power of the same number, the only way you can solve this (now) is by guess and check. If you didn't get an answer of ≈ 11.75, don't worry. You should be able to see the error on line 2. The exponent of 3x - 1 is on both the 2 and the 5. And both sides must have identical bases before you can set the exponents equal.

22. Horizontal: y = 0, vertical: x = 3
23. Horizontal: y = 1, vertical: x = 3
24. Horizontal: y = 0, vertical: x = 1, x = -1 **25.** Horizontal: y = 0
26. No, she will have approximately $7366.16 so she will need $633.84 more.
27. Losing 15% each year means the multiplier we use is .85. The equation is $1000(.85)^8 \approx \$272.49$. **28.** The height of the 10th bounce is given by $125(.97)^{10} \approx$ 92.18 feet. Still pretty high! **29.** The best way to do this problem is to choose an amount to be invested at each different level and see what happens. For instance, investing $1000 at 8% would be compounded yearly. Let's see what happens to this $1000 after 10 years under each of the different options. The first one will give $1000(1.08)^{10} \approx \$2158.92$. Under the second plan (compounded quarterly) we have $1000\left(1 + \frac{.07875}{4}\right)^{40} \approx \$2181.14$. Compounded monthly gives $1000(1 + \frac{.0775}{12})^{120} \approx \$2165.19$. And daily will yield $1000\left(1 + \frac{.075}{365}\right)^{3650} \approx$ $2116.84.

So it looks like quarterly is the way to go, with monthly being our second choice.
30. This is a function, it is continuous and nonlinear (curved). It has a y-intercept of (0, 5) and no x-intercepts. The domain is all x and the range is all y > 4. This has a horizontal asymptote of y = 4 and no vertical asymptotes. It is an exponential function.

SAT Prep

1. D **2.** D **3.** B **4.** A **5.** A **6.** 150 **7.** 19 **8.** n can be 3, 4, 5, or 6. **9.** 3.6 **10.** 5

Unit 4, More to Try

1. (-3, 0), (-1, 0), (0, 3) **2.** (-6, 0), (1, 0), (0, -6)
3. (4, 0), (-.5, 0), (0, -4) **4.** $(\frac{2}{3}, 0)$, (-4, 0), (0, 8) **5.** $(-\frac{5}{4}, 0)$, (1, 0), (0, -5)
6. $(\frac{2}{3}, 0)$, (0, 4) **7.** (-.8, 0), (.8, 0), (0, -25) **8.** ≈ (.926, 0), (-1.26, 0), (0, -7) **9.** ≈ (3.77, 0), (-.44, 0), (0, 5)
10. (2, 0), (0, -12) (Careful! It's a line!)
11. The formula starts with **-b**, the negative sign is left off. $\frac{-3 \pm \sqrt{6}}{3}$ **12.** Under the radical, **-4ac** should be **+ 40**. $\frac{-7 \pm \sqrt{89}}{-4}$ **13.** It should be **all over** 2a, not just the radical. $\frac{1 \pm \sqrt{7}}{4}$ **14.** It is a function.
15. It is not a function; it doesn't pass the vertical line test. **16.** It is not a function. For each input, x, there are two outputs (±).
17. It is a function.

18.

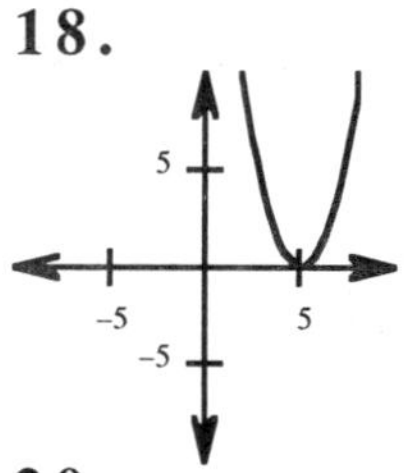

19.

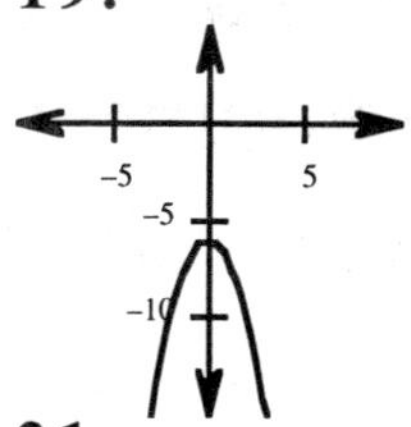

20.

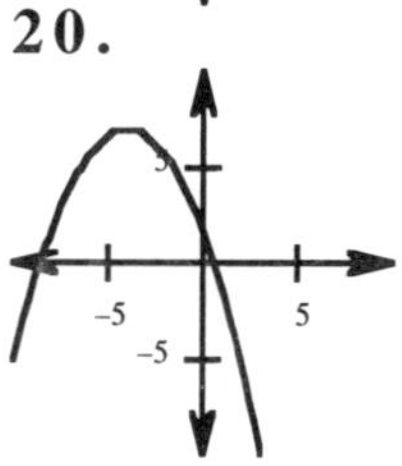

21.

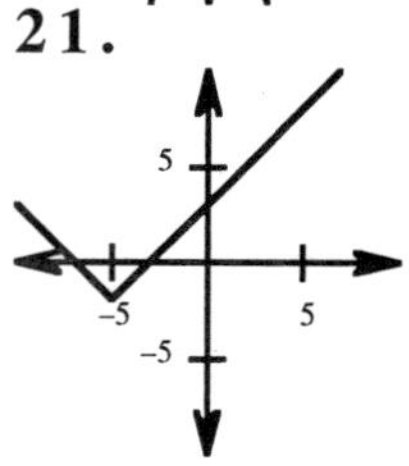

22.

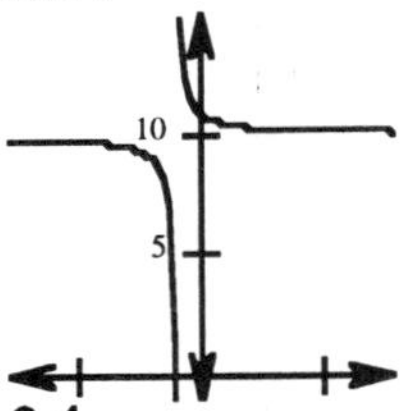

23.

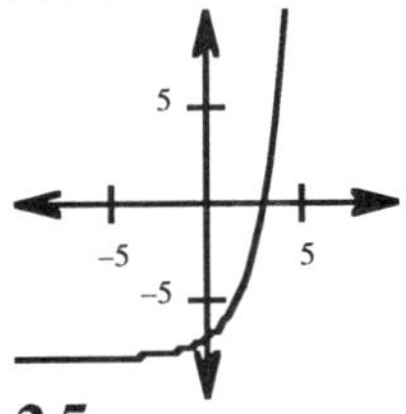

24.

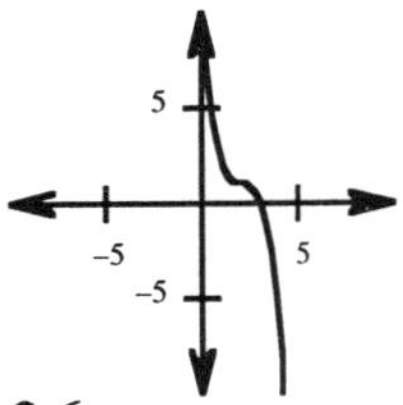

25.

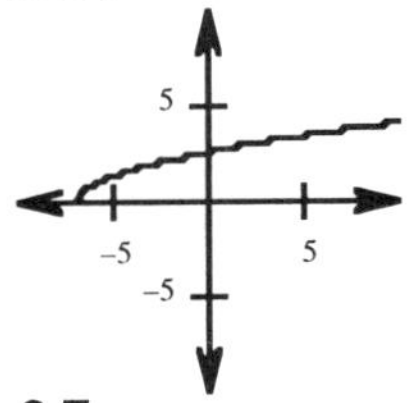

26.

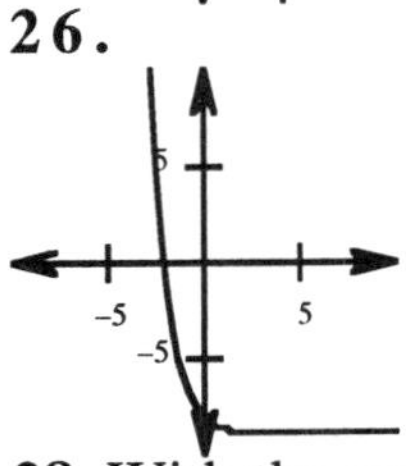

27.

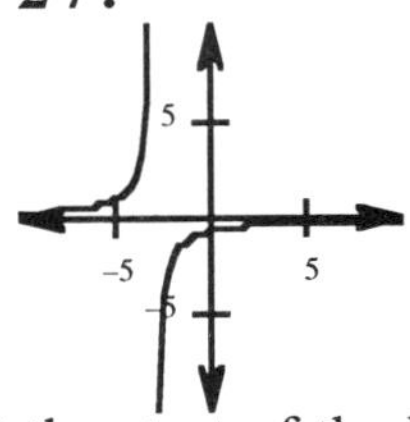

28. With the axes at the start of the kick, $y = -.05x(x - 100)$ will work. **29.** With axes at the start of the rise of sales, $y = -\frac{1}{6}x^2 + x$. Note: Let the x axis represent the number of months (then two points are (0, 0) and (6, 0)), and the y axis represent the number of <u>million</u> of units sold. The vertex is then (3, 1.5).
30. With the low point of the ramp (the vertex) at the origin, the equation $y = \frac{16}{75}x^2$ will work.

SAT Prep
1. D **2.** A **3.** D **4.** B **5.** B **6.** C
7. 250 **8.** 5 **9.** 20 **10.** 1.5

Unit 5, More to Try
1. (6, 2), the two lines cross at this point.
2. (8, 3), the two lines cross at this point.
3. (-1, 8), the two lines cross at this point.
4. $\left(\frac{-226}{11}, \frac{185}{11}\right)$, these two lines cross at this point. This one is not so hard to check if you use your graphing calculator. Leave your answer as a fraction if you do. **5.** No solution. These two lines are parallel -- they never cross. This is called **inconsistent.** **6.** (3, 1, -4), these three planes intersect in a point.
7. (1, -5, -1), these three planes intersect in a point. **8.** (3, 1, 2.5), these three planes intersect in a point. **9.** No solution or inconsistent. Two of these planes are parallel. **10.** (36, 24, 12), these three planes intersect in a point.

11.

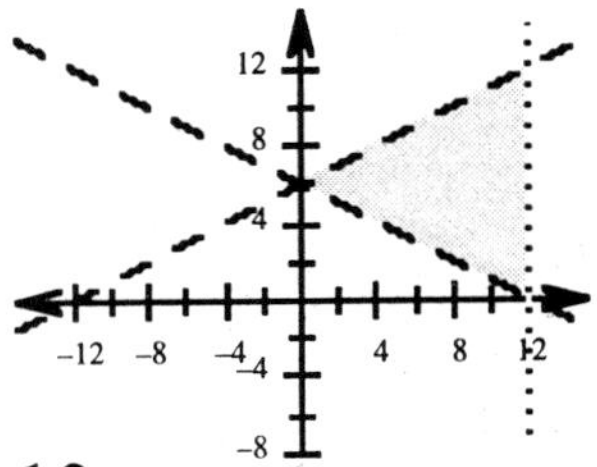

12.

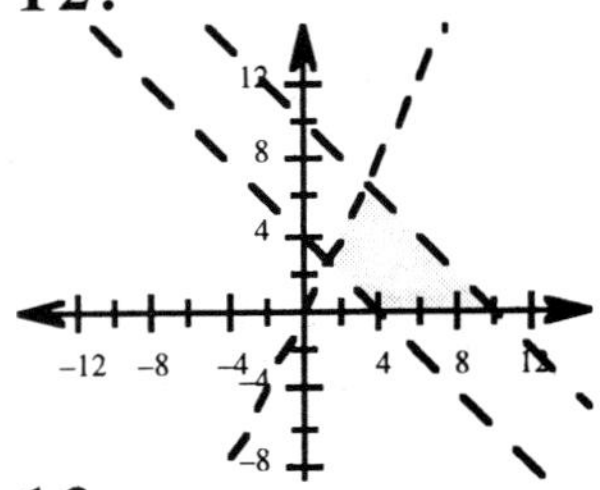

13.

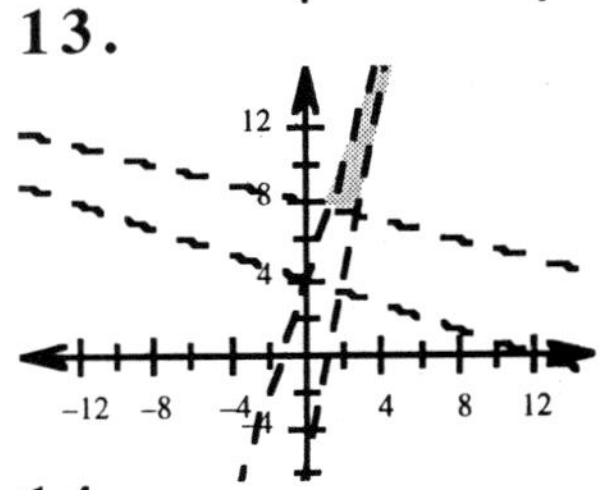

14.

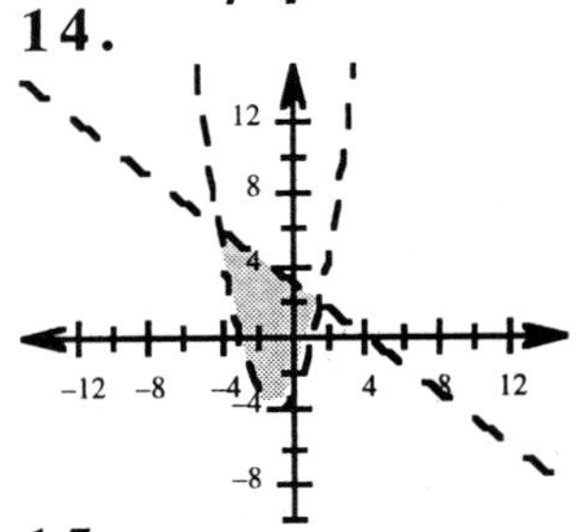

15.

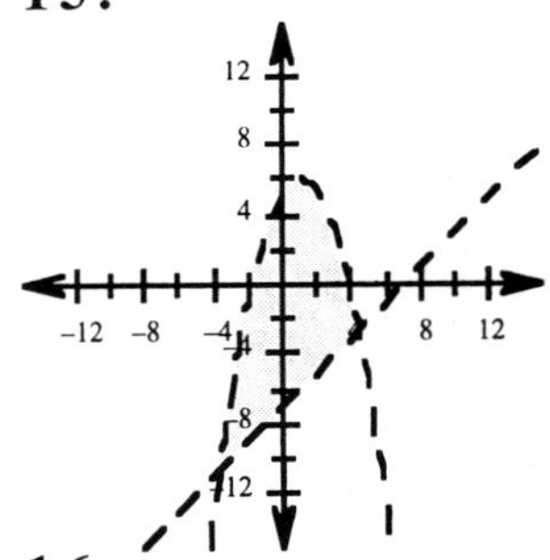

16.

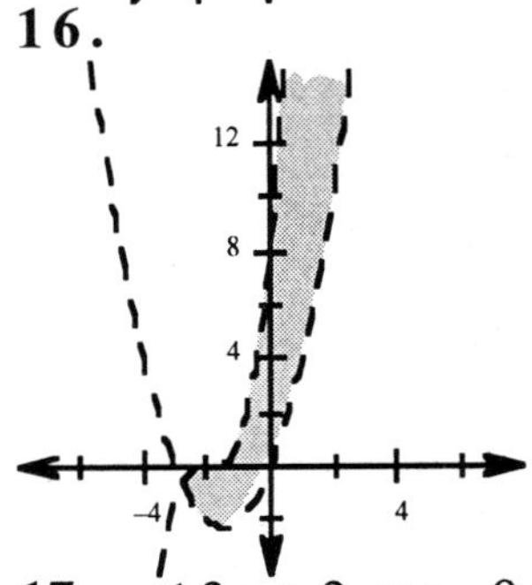

17. $y \le 3x + 2$, $y > -0.5x + 6$, $x < 4$
18. $y < -x^2 + 10$, $y \ge x + 1$

19. $y = -3x^2 + 5x - 10$ **20.** The equation which fits this data is $y = .04x^2 - 3.6x + 100$. This model says that an 80 year old person would make 68 misunderstandings in a 24 hour period. A one year old would make ≈ 96. The age that understands the most would be the age at which the number of misinterpretations is the lowest. This is at the vertex of this function. The vertex is at (45, 19) so 45 year olds have the lowest number with only 19 mistakes. **21.** With rounding, $y = -.31x^2 + .43x + 5.4$.
22. The graph of the feasibility region is

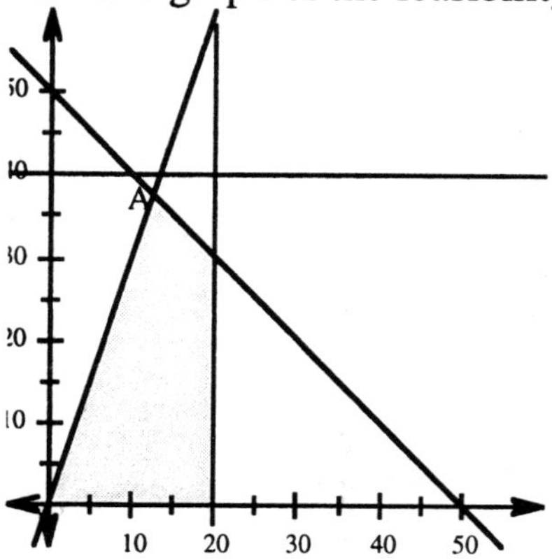

The inequalities are $x + y \le 50$, $x \le 20$, $y \le 40$ and $y \le 3x$, where x is the number of Hauteans and y is the number of Zotions. The profit is given by $P = x + 1.5y$. If you use the profit line and slide it up to find the maximum profit, you would find point A (12.5, 37.5) to maximize, but there is a problem with this point. Davis and Tess can not make a half of an alien (or at least that doesn't seem possible). Try these nearby points: (12, 37), (12, 38), (13, 37), and (13, 38). Which gives maximum profit **and** is still in the feasibility region? Just (13, 37). They should make 13 Hauteans and 37 Zotions for a profit of $68.50.

23. $\begin{bmatrix} 18 & 9 & 9 \\ 48 & 63 & 42 \end{bmatrix}$, cost: $106146, sell for $202803, profit: $96657.

SAT Prep

1. E **2.** B **3.** D **4.** E **5.** A **6.** 10 **7.** 280 **8.** 9 **9.** 26 **10.** 45

Unit 6, More to Try

1. $y = \frac{x}{8} + 13$ **2.** $y = -\frac{4}{3}x + 8$ **3.** $y = \frac{3}{5}x - 2$ **4.** $y = \pm\sqrt{x - 6}$. Did you remember the ±? **5.** $y = \frac{3}{x - 6}$ **6.** $y = \frac{5}{x}$ No, this isn't a mistake. This function is its own inverse! Graph it if you don't believe it! **7.** $y = -1 \pm \sqrt{x + 3}$ **8.** $y = -2 + \sqrt[3]{x}$ **9.**

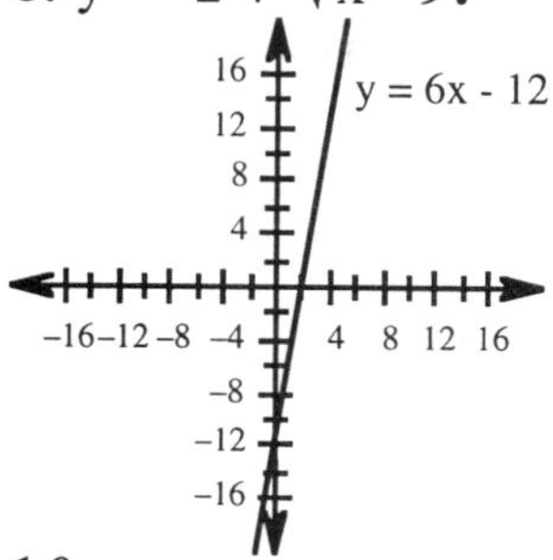

10.

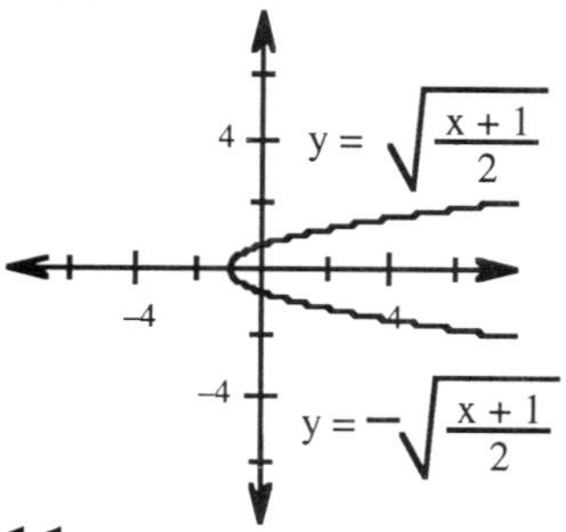

11.

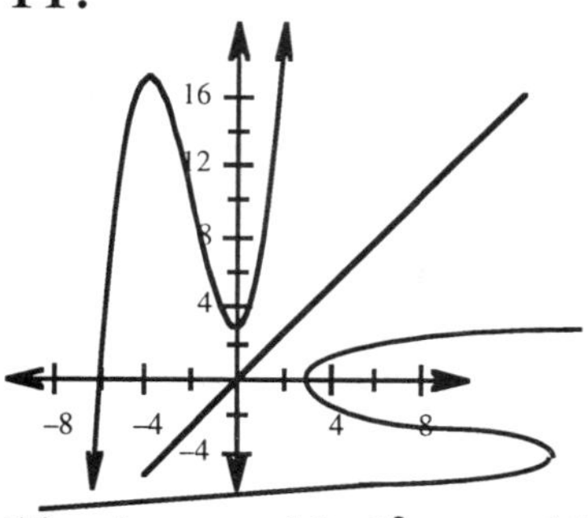

12. $4^y = x$ **13.** $2^3 = x$ **14.** $5^x = 30$ **15.** $\log_4 80 = x$ **16.** $\log_{.5} 64 = x$ (Just for fun and for practice, you may want to try to solve these equations, when possible. #12 has both an x and y so you can't solve that one without another equation. But, here are the answers to the rest if you care to try: 13: 8, 14: ≈ 2.113, 15: ≈ 3.160, 16: -6.)
17. $x \approx 2.336$ **18.** $x \approx .721$ **19.** $x = 2$. Rewrite as an exponential equation and then use logs (base 10). **20.** $x = 81$
21. $x \approx 1.119$. Did you divide by 5 first?
22. Careful! No logs here! $x \approx 1.596$
23. Rewrite as an exponential equation and then take the fourth root of both sides. $x \approx 3.162$ **24.** $x \approx 2.365$ **25.** $y = 7(3^x)$
26. $y = -1(-5.667)^x + 8$ **27.** Use the general equation $y = km^x + c$ with the data points (0, 88) -- that's letting 1:00 am be the starting time of 0 -- and (1, 84) -- that's

one hour later. Then your equation should become $y = 14(.6429)^x + 74$. Set equal to 98.6 (body temperature when alive) and solve for x. $x \approx -1.276$. Time of death is approximately 1 hour 17 minutes before 1:00 am, or 11:43 pm. **28.** Use the equation $y = km^x$ with data points (0, 1.75) and (22, 6.50). The equation is $y = 1.75(1.06146)^x$, let x = 29 and solve for y. Cost ≈ $9.87.

SAT Prep
1. D **2.** B **3.** D **4.** C **5.** C **6.** A
7. D **8.** A **9.** A **10.** C

Unit 7, More to Try
1. Yes, degree 7. **2.** No, we can have a term like $\frac{5}{x}$ because that is the same as $5x^{-1}$. Polynomials only have positive integer powers on the variable. **3.** Careful! Multiply this one out and you will get $6x^2 + 13x + 2$, which is a polynomial, degree 2.
4.

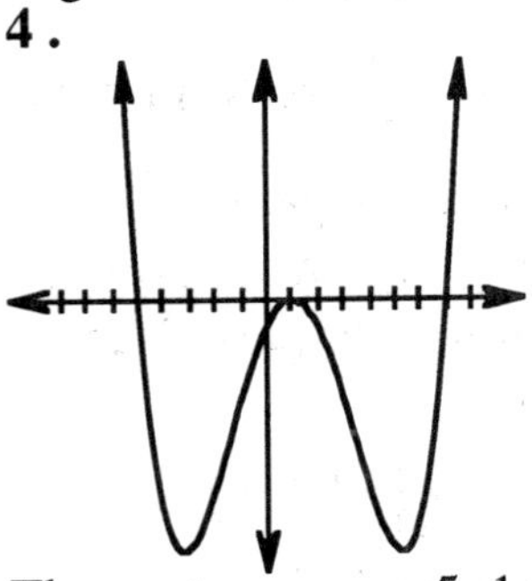

The roots are x = -5, 1 and 7 with x = 1 being a double root. Remember a double root is where the graph is tangent. This has a positive orientation.
5.

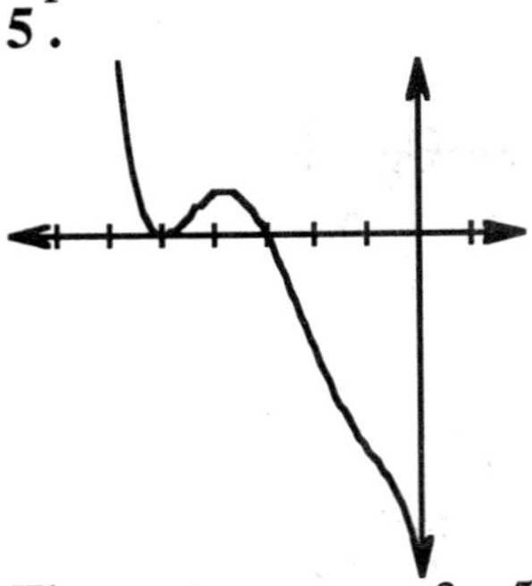

The roots are x = -3, -5 (which is a double root) and then two complex roots (from the factor $x^2 + 2$: $\pm \mathbf{i}\sqrt{2}$). Since x = -5 is a double root, this is the five roots we would expect. The extra "squiggle" in the curve is because of the complex roots. Your curve should have the negative orientation, but if you put a squiggle elsewhere that is find.

6.

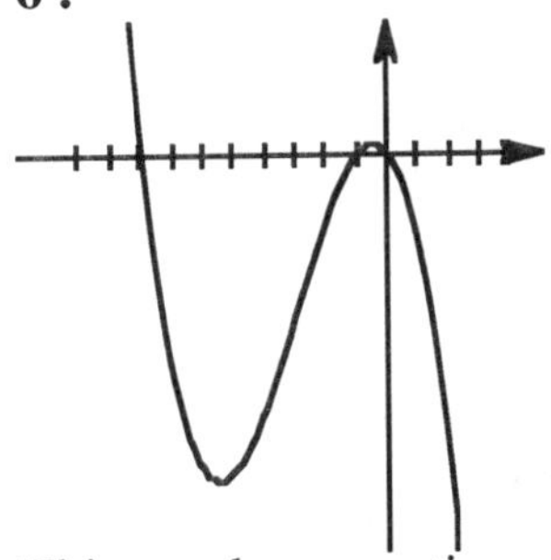

This one has negative orientation and x = 0 is a root don't forget!
7.

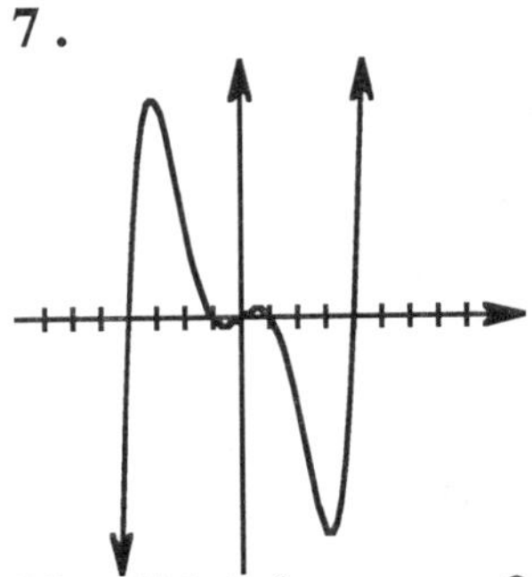

You didn't forget x = 0 as one of the roots this time, right? Also, $x^2 - 1$ gives us two roots. This term is just not factored yet: (x + 1)(x - 1). **8.** $x \approx .79729$
9. $x \approx -1.949$, $x \approx 2.9025$, $x \approx 17.688$, Yes! There are three. Zoom out to see that third one. **10.** $x \approx 2.3803$, $x \approx 5.5031$, and yes, there is one very, **very** close to zero. In fact, it is zero if we are only accurate to three decimal places. Through repeated guess and checking I found that $x \approx 10^{-13}$ is a good estimate! Don't worry if you didn't get this third one. Just trying to keep you on your toes! **11.** 4 + 5**i**
12. 4**i** **13.** 84 **14.** 11 + 29**i** **15.** 13
16. 28 **17.** (4, -3) and (-4, -3).

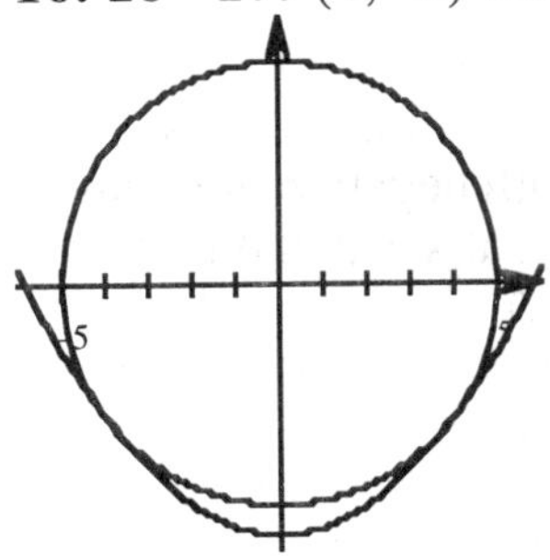

18. (6, 3) and (-1, 17)

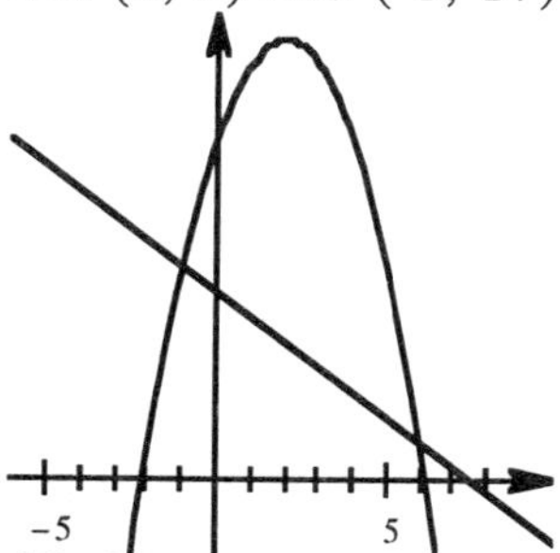

19. No solution.

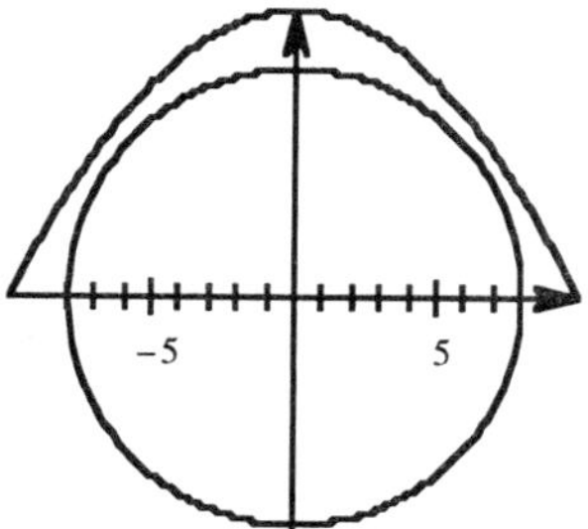

20. $(2 + \sqrt{14},\ 2 - \sqrt{14})$ and $(2 - \sqrt{14},\ 2 + \sqrt{14})$

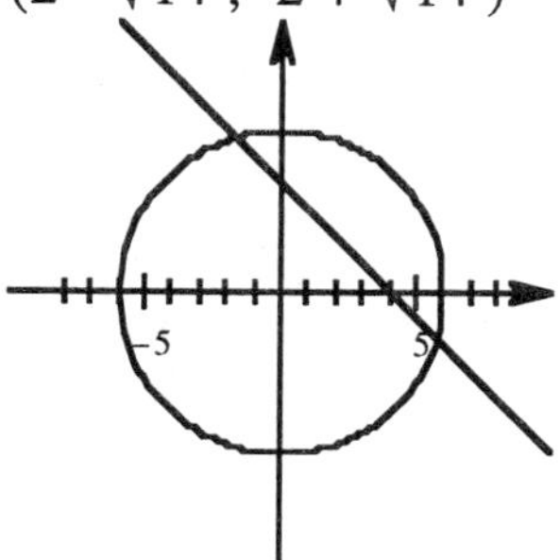

21. A cubic with one real root at x = 0 and one double root at x = -4. Positive orientation. **22.** A fourth degree polynomial with real roots at x = -5, x = -3 and a double root at x = 5. This has negative orientation. **23.** A fifth degree polynomial with five real roots: x = -5, x = -1, x = 2, x = 4, and x = 6. This has positive orientation. **24.** A third degree polynomial with negative orientation and with one real root at x = 5 and two complex roots. **25.** A fourth degree polynomial with positive orientation and two real roots: x = -6 and x ≈ -.5 and two complex roots. **26.** A fifth degree polynomial with negative orientation and with one real root at x = -4 and four complex roots. **27.** $y = (x + 3)(x - 1)(x - 4)$
28. $y = -.1(x + 5)(x + 2)(x - 3)(x - 5)$
29. $y = \frac{1}{12}(x + 3)^2(x - 1)(x - 4)$
30. $y = -\frac{1}{18}(x^2 - 3x - 10)(x^2 + 9)$

SAT Prep
1. E **2.** A **3.** C **4.** B **5.** D **6.** .1
7. $\frac{7}{4}$ **8.** 21 **9.** 6 **10.** 360

Unit 8, More to Try
1.

Amplitude is 2, period is π.

2.

Amplitude 0.5, period 2π **3.** The graph shows one cycle already. Amplitude is 3 and period is 4π. **4.**

, amplitude is 2.5, period is $\frac{\pi}{3}$.

5. Amplitude: 2, period: $\frac{2\pi}{3}$ **6.** Amplitude: $\frac{1}{2}$, period: 2π **7.** Amplitude: 3, period: $\frac{\pi}{2}$
8. Amplitude: 1, period: 6π **9.** Amplitude: 1, period: 2π **10.** Amplitude: 5, period: 2π.

11.

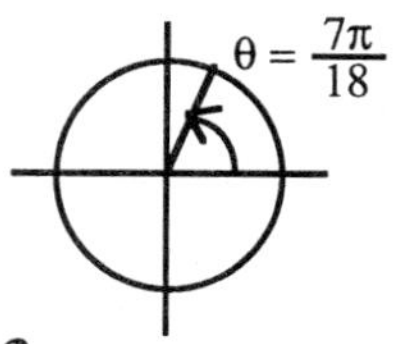

12.

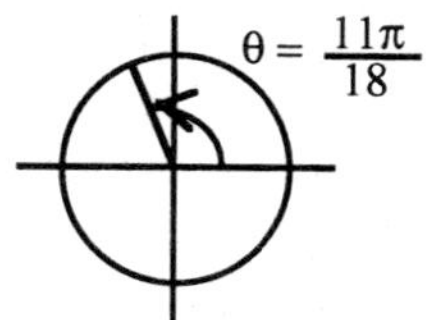

13.

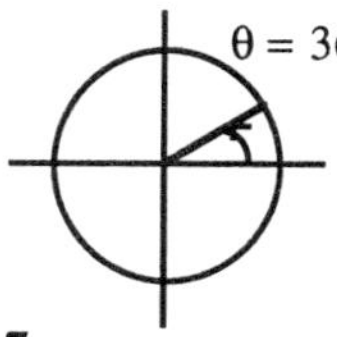

14.

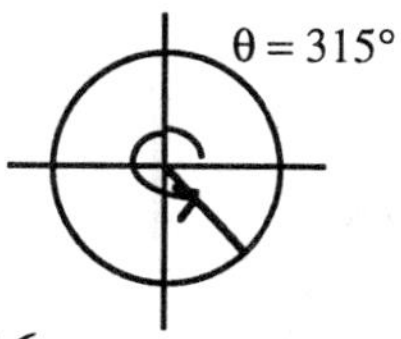

15.

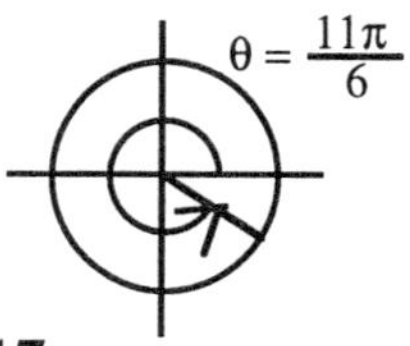

16.

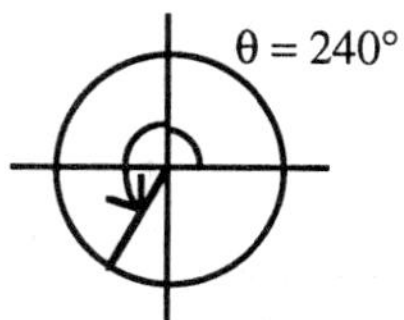

17.

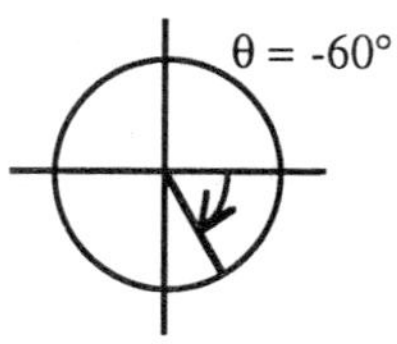

18.

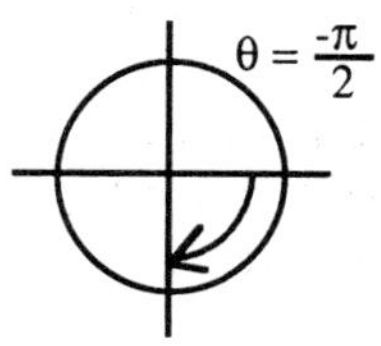

19.

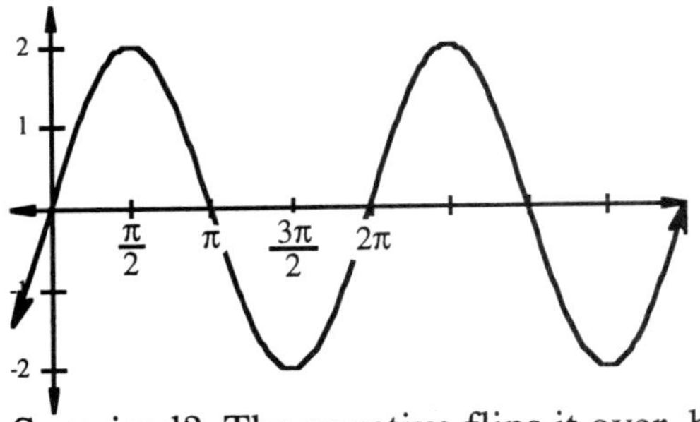

Surprised? The negative flips it over, but the "+ π" shifts it right back to how it looks originally!

20.

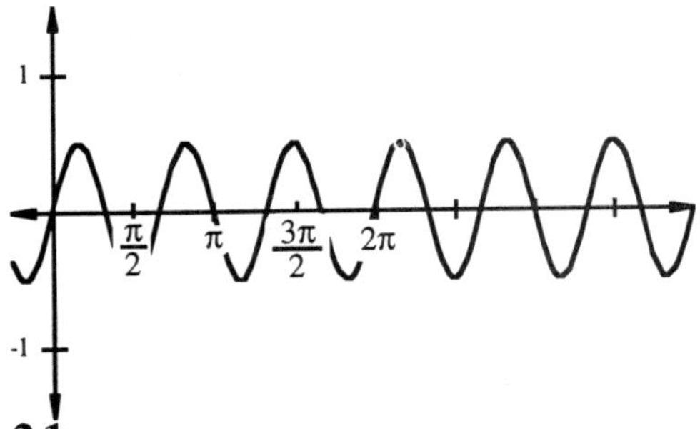

21.

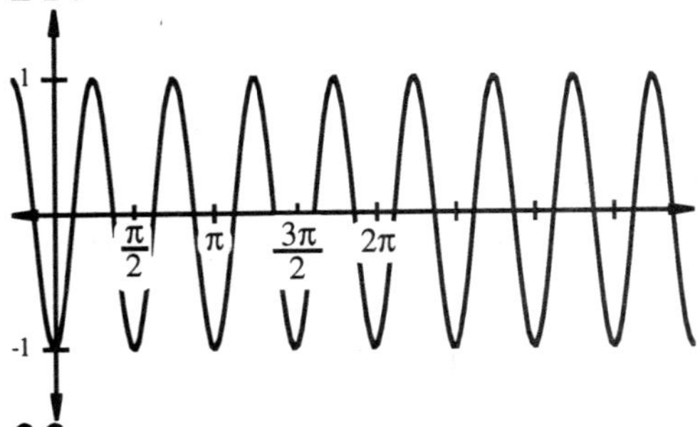

22.

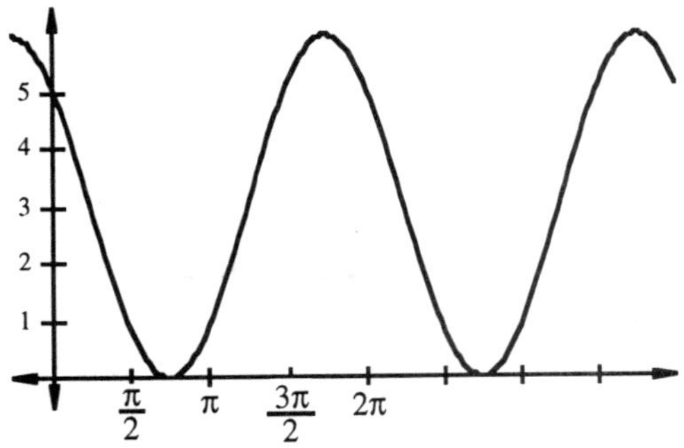

23.

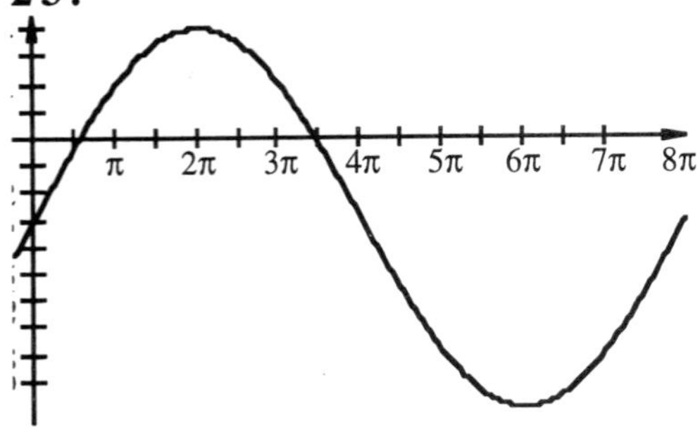

24. $y = -10\cos\left(\frac{1}{10}x\right) + 8$, and there are other possible equations which will work.
25. $y = -6\cos\left(\frac{2\pi}{3}x\right)$ works if we let the graph be symmetric about the x-axis. The x-axis does not have to represent the ground. If you let the x-axis represent the ground, you equation might look like:
$y = -6\cos\left(\frac{2\pi}{3}x\right) + 14$.

SAT Prep
1. B **2.** A **3.** A **4.** B **5.** A

Unit 9, More to Try
1. 23.32 **2.** 16.46 (Throw out the negative answer.) **3.** 11.04. Remember you can easily find m∠A because the angles of a triangle sum to 180°. **4.** 45.14 **5.** 9.34 **6.** Aha! The ambiguous case! c = 5.32 or 1.32. **7.** 8.92 **8.** 61.28° **9.** 16. Did you realize that this is just a 30°-60°-90° triangle? **10.** Oh no! The triangle inequality doesn't hold. There is no triangle! **11.** 53.13° **12.** 6.88 **13.** 39.03 **14.** 71.38° **15.** 4.32 **16.** 10.97 sq un. **17.** 785.35 sq. un. **18.** 3.42 sq. un. **19.** Use Law of Cosines first to get m∠C ≈ 132.18°. Then A ≈ 12.97 sq. un. **20.** x = 0° **21.** x = 60° **22.** x ≈ 143.13° or -36.87°, and 36.87° . **23.** Did you know that the only way for a fraction to equal zero is for the numerator to equal zero? It is true! Set the top equal to zero and solve. x ≈ 138.59°. **24.** It is a function, it is not continuous. It has asymptotes every 180° (or every π radians), so the domain is all x except 0°, 180°, 360°, ... or 180k for all integer k. The range is y ≥ 1 and y ≤ -1. No x-intercepts and no y-intercepts. A "vertex" (it isn't really a vertex -- it is a maximum point or minimum point for each section of the graph) occurs at 90°, 270°, ... or 90k for all integer k ≠ 0. It is cyclic, periodic, with period 2π. No amplitude because no "high" and "low" point. Can you think of anything else? **25.** These produce the same graph so we can conclude that $\cos 2x = 2\cos^2 x - 1$.

SAT Prep
1. 57 **2.** 16 **3.** $\frac{1}{4}$ **4.** 19 **5.** 12

Unit 10, More to Try

1. $y = (x - 4)^2 + 9$, vertex (4, 9)
2. $y = (x + 4)^2 - 6$, vertex (-4, -6)
3. $y = 3(x - 1)^2 + 12$, vertex (1, 12)
4. $y = -4(x + 3)^2 - 2$, vertex (-3, -2)
5. $y = \frac{1}{2}(x + 6)^2 + 7$, vertex (-6, 7)
6. $y = .1(x + 7)^2 + 5$, vertex (-7, 5)
7. $y = 3(x - 9)^2 - 6$, vertex (9, -6)
8. $y = -4\left(x - \frac{5}{2}\right)^2 + \frac{7}{2}$, vertex $\left(\frac{5}{2}, \frac{7}{2}\right)$
9.

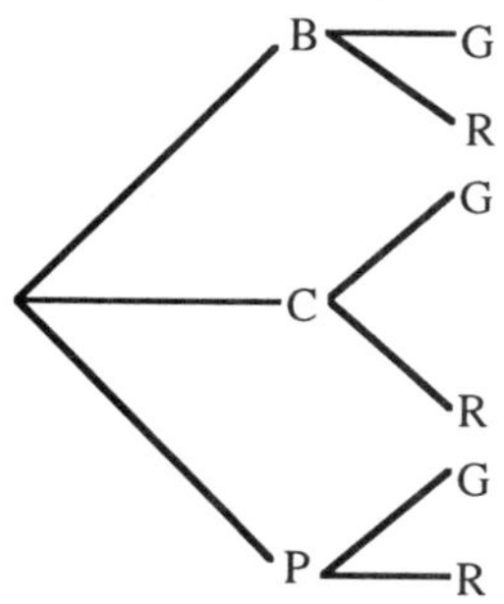

10. $\frac{1}{6}$ **11.** $\frac{1}{6}$ **12.** $\frac{1}{3}$ **13.** $\frac{15}{36} = \frac{5}{12}$ **14.** $\frac{12}{36} = \frac{1}{3}$ **15.** $\frac{35}{36}$ **16.** 30% or $\frac{3}{10}$
17. 0.34

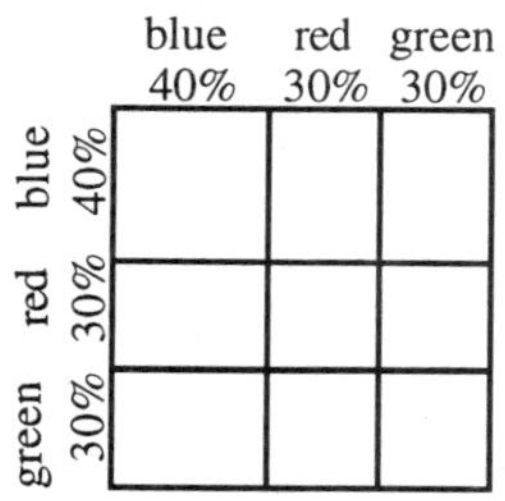

18. ≈ .26 **19.** .25
20. (a) $\frac{1}{3}$ (b) $\frac{2}{9}$ (c) $\frac{4}{9}$ **21.** ≈ 1.11 **22.** .63 **23.** .18 **24.** .12 **25.** .84 **26.** 4324 **27.** 3996 **28.** Levels off to 50-50, 4000 of each kind.

SAT Prep

1. A **2.** E **3.** D **4.** E **5.** A

Unit 11, More to Try.

1. 3628800 **2.** 604800 **3.** 38955840 **4.** 88 **5.** 5112 **6.** 43680 **7.** 56 **8.** 1 **9.** 362880 **10.** 495 **11.** 1 **12.** 1 **13.** 5! = 120 **14.** 2(4!) = 48 **15.** (26)(26)(26)(10)(10)(10) = 17576000 **16.** 2300 **17.** 13800 (On a cone, order matters -- at least it does to me! I like my favorite on the bottom!) **18.** ${}_{52}C_5$ = 2598960 **19.** This is tricky and tough! There are 13 different "types" of cards: twos, threes, fours, ..., Jacks, Queens, Kings and Aces. We need to choose which of the 13 we want three of $({}_{13}C_1)$. Once we choose what type (say we pick Jacks) then we need to choose which three out of the four to take $({}_4C_3)$. Then from the remaining 12 types, we choose which type to have two of $({}_{12}C_1)$. Then again we need to choose which two out of the four $({}_4C_2)$. This gives us ${}_{13}C_1 \cdot {}_4C_3 \cdot {}_{12}C_1 \cdot {}_4C_2 = 3744$. **20.** We have already done the tough stuff: $\frac{3744}{2598960}$. **21.** Each time we are reaching in and pulling out 8 marbles -- order does not matter. The number of ways to do this is ${}_{36}C_8$. This is the number in the sample space, i.e. the denominator. How many ways can we pull out all blue? ${}_{12}C_8$. Therefore the probability is $\frac{{}_{12}C_8}{{}_{36}C_8} \approx$ .0000164. **22.** Same denominator. Now we want to choose 4 from the 12 blue, ${}_{12}C_4$, and 4 from the 4 whites, ${}_4C_4$. $\frac{{}_{12}C_4 \cdot {}_4C_4}{{}_{36}C_8} \approx$ .0000164 -- Yes! The same answer! **23.** Seven green: ${}_7C_7$, one yellow: ${}_5C_1$. $\frac{{}_7C_7 \cdot {}_5C_1}{{}_{36}C_8} \approx$ .0000001652 **24.** Here we have to get at least one red: ${}_8C_1$, and at least two yellow: ${}_5C_2$, but the other five marbles can come from the rest of the pot: ${}_{33}C_5$. Therefore, $\frac{{}_8C_1 \cdot {}_5C_2 \cdot {}_{33}C_5}{{}_{36}C_8} \approx$ .627 **25.** To get no blue marbles means we want all eight from the other 24 non-blue marbles. $\frac{{}_{24}C_8}{{}_{36}C_8} \approx$ .0243.

SAT Prep

1. $\frac{1}{729}$ **2.** $\frac{43}{6}$ **3.** 0 **4.** 228 **5.** 24.5

Unit 12, More to Try

1. 63 **2.** ≈ 7.998 **3.** ≈ 597.3 **4.** ≈ 133.35 **5.** 600 **6.** -1.254, 1.5202, 1.1453, -2.075, .62036, -.4295, -.3545, .17043, -.2045, .0245, .84532 **7.** Since the median is 600 (a high SAT score) and half of this class is above this, they are probably pretty good!

8.

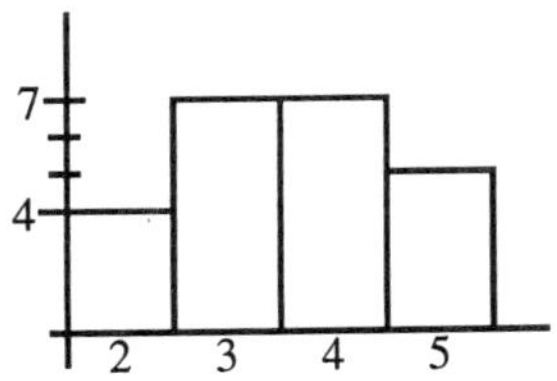

9. mean = 3.56, median = 4. The histogram shows it nicely. It is distributed fairly normally and evenly. **10.** Biased. We would expect people living in D.C. to be more familiar with federal government issues than in other parts of the country. **11.** Unbiased although it is a small sample. **12.** Unbiased.

SAT Prep

1. A **2.** E **3.** B **4.** C **5.** 2 and 8 **6.** 112.5 miles **7.** 20 **8.** Any of 2, 3, 4, or 5 will work. **9.** $1 - \frac{\pi}{4} \approx .2146$ **10.** 19, 9, 11, -19, -9, -11

Unit 13, More to Try

1. 504 **2.** 21845 **3.** 4095.9375 **4.** -330 **5.** 40200 **6.** 243 **7.** For n = 1, $2(1) = 2 \le 2^1$, check. Assume it true for n = k: $2(k) \le 2^k$. Prove it true for n = k + 1: $2(k + 1) = 2k + 2 \le 2^k + 2$, by the assumption that n = k is true. But, $2 \le 2^k$, so we can continue the string of inequalities: $2(k + 1) = 2k + 2 \le 2^k + 2 \le 2^k + 2^k = 2(2^k) = 2^{k+1}$. Ta Da! **8.** For n = 1 we have $\frac{1}{2} = 1 - \frac{1}{2^1} = \frac{1}{2}$. Check! Assume the statement is true for n = k, that is: $\frac{1}{2^1} + \frac{1}{2^2} + \frac{1}{2^3} + \frac{1}{2^4} + ... + \frac{1}{2^k} = 1 - \frac{1}{2^k}$.
Prove it true for n = k + 1:

$$\frac{1}{2^1} + \frac{1}{2^2} + \frac{1}{2^3} + \frac{1}{2^4} + ... + \frac{1}{2^k} + \frac{1}{2^{k+1}}$$

$$= (1 - \frac{1}{2^k}) + \frac{1}{2^{k+1}} = 1 - \frac{1}{2^k} + \frac{1}{2^{k+1}}$$

$$= 1 - \frac{1}{2^k} \cdot \frac{2}{2} + \frac{1}{2^{k+1}}$$

$$= 1 - \frac{2}{2^{k+1}} + \frac{1}{2^{k+1}}$$

$$= 1 - \frac{1}{2^{k+1}}.$$ Ta Da! **9.** Note: since the proof asks for $n \ge 3$, we will not start by checking n = 1. We start by checking n = 3: $2^3 = 8 \ge 2(3) + 1 = 7$. Assume it true for n = k: $2^k \ge 2k + 1$. Prove it true for n = k + 1.
$2^{k+1} = 2^k \bullet 2 \ge 2(2k + 1)$
$= (2k + 1) + (2k + 1) \ge 2k + 3 = 2k + 2 + 1$
$= 2(k + 1) + 1$. Ta Da!

10. $\frac{(x + 2)^2}{16} + \frac{(y + 1)^2}{4} = 1$, ellipse, center (-2, -1), foci $(-2 \pm \sqrt{12}, -1)$ $e = \frac{\sqrt{3}}{2}$
11. $\frac{(y - 6)^2}{25} + \frac{(x + 3)^2}{4} = 1$, hyperbola, center (-3, 6), "vertices": (-3, 11) and (-3, -1), foci: $(-3, 11 + \sqrt{29})$, $(-3, -1 - \sqrt{29})$, asymptotes: slopes: $\pm\frac{5}{2}$ so equations: $y = \frac{5}{2}x + \frac{27}{2}$ and $y = -\frac{5}{2}x - \frac{3}{2}$, $e = \frac{\sqrt{29}}{5}$ **12.** $y = 3(x - 2)^2 + 7$, parabola, vertex: (2, 7), focus $(2, 7 + \frac{1}{12})$, e = 1 (All parabolas have eccentricity of 1!) **13.** $(x - 6)^2 + (y + 5)^2 = 169$, center (6, -5), radius 13, e = 0.

SAT Prep

1. D **2.** D **3.** D **4.** 4 **5.** 6 **6.** 7 **7.** 8 **8.** 9 g **9.** 20 **10.** $\frac{1}{18}$